PRAISE FOR MICHAEL BARTON

From "*The Power of Regenerative Agriculture:*"

This is your go-to book to help inspire and educate yourself about the possibility of how you can play a role in the interconnection of farming systems with the entirety of the ecological system.

— D. G. FARNSWORTH

Well done book. I liked the chapter summaries at the end of each chapter. Chapter 6, was spot on in describing the challenges and barriers to implementing regenerative agriculture. The author has a passion for this subject and has provided very clear and concise data. The Afterword ends the book on a positive note and a view of hope that we'll get there.

— SRP

These guides to being environmentally responsible often come across as a little preachy. This is not one of those. It's informative and well-written, with a cheerful attitude toward ecologically friendly farming methods.

— LWHIT

RAINWATER HARVESTING AND NO-TILL FARMING

RAINWATER HARVESTING AND NO-TILL FARMING

SUSTAINABLE WATER CONSERVATION METHODS AND ENVIRONMENTALLY FRIENDLY PRACTICES FOR YOUR HOME (2-IN-1 COLLECTION)

SUSTAINABLE LIVING AT HOME

MICHAEL BARTON

Book
Bound Studios

The information contained in this book is based on the author's personal experiences and research. While every effort has been made to ensure the accuracy of the information presented, the author and publisher cannot be held responsible for any errors or omissions.

This book is intended for general informational purposes only and is not a substitute for professional medical, legal, or financial advice. If you have specific questions about any medical, legal, or financial matters, you should consult with a qualified healthcare professional, attorney, or financial advisor.

Book Bound Studios is not affiliated with any product or vendor mentioned in this book. The views expressed in this book are those of the author and do not necessarily reflect the views of Book Bound Studios.

To all the individuals who are committed to creating a more sustainable and environmentally responsible future. May this book inspire and guide you as you take important steps towards reducing your environmental footprint and contributing to a healthier planet. Your dedication and efforts are crucial in ensuring a brighter future for generations to come.

Nature provides a free lunch, but only if we control our appetites.

— WILLIAM RUCKELSHAUS

CONTENTS

From the Author xiii
Introduction xv

RAINWATER HARVESTING FOR YOUR HOME

Introduction 3
1. The Importance of Rainwater Harvesting for Your Home 13
2. Understanding the Rainwater Harvesting Process 23
3. Assessing Your Home's Rainwater Harvesting Potential 33
4. Designing and Planning Your Rainwater Harvesting System 43
5. Choosing the Right Rainwater Collection and Storage Equipment 55
6. Installing Your Rainwater Harvesting System 67
7. Maintaining and Monitoring Your Rainwater Harvesting System 73
8. Utilizing Harvested Rainwater for Various Household Purposes 83
9. Legal and Regulatory Considerations for Rainwater Harvesting 93
10. The Environmental and Economic Benefits of Rainwater Harvesting 103
Afterword 113

NO-TILL FARMING FOR YOUR HOME

Introduction 123
1. The History and Evolution of No-Till Farming 133
2. Understanding Soil Health and Its Importance 143
3. The Science Behind No-Till Farming Techniques 153
4. Preparing Your Home Garden for No-Till Farming 165
5. Selecting the Right Crops for Your No-Till Garden 179

6. Implementing Cover Crops and Crop Rotation Strategies 189
7. Managing Weeds and Pests in a No-Till Garden 201
8. Fertilization and Nutrient Management in No-Till Farming 213
9. Harvesting and Post-Harvest Practices for No-Till Gardens 223
10. The Environmental and Economic Benefits of No-Till Farming at Home 235
Embracing a Sustainable Future with No-Till Farming 245

Epilogue 253

About the Author 255
From the Author 257

INTRODUCTION

Welcome to *"Rainwater Harvesting and No-Till Farming."* In today's world, where climate change and environmental degradation are becoming increasingly pressing issues, adopting eco-friendly practices in our daily lives is more important than ever. That's where this book comes in - it offers practical guidance on two essential practices that can help reduce your environmental footprint and contribute to a healthier planet: rainwater harvesting and no-till farming.

In the first part of this 2-in-1 collection, *"Rainwater Harvesting for Your Home,"* you'll learn about the importance of rainwater harvesting for sustainable living and how to use this practice to provide a reliable water source for your home. You'll explore the essential components of a rainwater harvesting system, assess your home's potential for rainwater harvesting, and navigate the legal and regulatory considerations to ensure compliance with local laws and regulations.

In the book's second part, *"No-Till Farming for Your Home,"* you'll discover the rich history and evolution of no-till farming and its crucial role in sustainable agriculture. You'll learn about the numerous benefits of no-till practices, from improved soil health to increased crop yields and reduced labor. You'll also master essential tools and techniques for successful no-till gardening. Along the way, you'll overcome common challenges no-till gardeners face with practical solutions and expert advice.

With this 2-in-1 collection, you'll have everything you need to transform your home and garden into a sustainable haven. Whether you're a seasoned gardener or just starting, this book offers a clear and concise guide to help you take one more step toward a greener future. So, what are you waiting for? Start reading now and join the global movement toward a more sustainable and eco-friendly future.

RAINWATER HARVESTING FOR YOUR HOME

A COMPREHENSIVE GUIDE TO SUSTAINABLE RAINWATER COLLECTION FOR HOMEOWNERS

INTRODUCTION

Welcome to *"Rainwater Harvesting for Your Home,"* where we'll explore the fascinating world of harvesting rainwater! As we become more aware of our environmental impact, we must find sustainable living ways. One way we can do this is by **harvesting rainwater.**

In this chapter, we'll dive into the importance of rainwater harvesting and how it can contribute to a more sustainable lifestyle. We'll cover the essential components of a rainwater harvesting system, so you can understand how it all works. Then, you'll learn how to assess your home's potential for rainwater harvesting, so you can make informed decisions about whether or not it's right for you.

We'll also explore legal and regulatory considerations for rainwater harvesting, so you can be sure you're following the rules and regulations in your area. Finally, we'll discuss the many benefits of rainwater harvesting and how it can help us move toward a greener future.

So, whether you're a seasoned environmentalist or just starting your journey toward sustainable living, this book has something for you. Let's get started!

The Importance of Rainwater Harvesting for Sustainable Living

In today's rapidly changing world, sustainable living has become more critical. As we strive to reduce our ecological footprint and conserve our precious natural resources, rainwater harvesting emerges as a practical and effective solution. This section will delve into the significance of rainwater harvesting and its role in promoting a sustainable lifestyle.

Rainwater harvesting, at its core, is collecting, storing, and utilizing rainwater for various purposes, such as irrigation, watering plants, flushing toilets, and even drinking when adequately treated. By capturing and reusing this valuable resource, we can significantly reduce reliance on traditional water sources, such as municipal water supplies, wells, and rivers. This, in turn, has numerous environmental, economic, and social benefits.

First and foremost, rainwater harvesting helps conserve water, one of our most vital natural resources. With climate change causing unpredictable weather patterns and increasing instances of drought, water scarcity has become a pressing global issue. By harvesting rainwater, we can alleviate the pressure on our existing water sources and ensure a more sustainable supply for future generations.

Moreover, rainwater harvesting reduces the demand for our overburdened stormwater management systems. By capturing rainwater before it enters the storm drains, we can minimize the risk of urban flooding and decrease the pollutants that flow into our rivers, lakes, and oceans. This protects our aquatic ecosystems and improves the overall water quality for humans and wildlife.

From an economic standpoint, rainwater harvesting can lead to significant cost savings for homeowners. By supplementing or replacing the need for municipal water, households can reduce their monthly water bills and become more self-sufficient. Additionally, implementing rainwater harvesting systems can increase property values, as potential buyers see them as an attractive and eco-friendly feature.

Rainwater harvesting also plays a crucial role in promoting social equity. In many parts of the world, clean water is a luxury that millions cannot afford. By implementing rainwater harvesting systems in these communities, we can help bridge the gap between the water-rich and water-poor, ensuring that everyone has the opportunity to live a healthy and dignified life.

Lastly, rainwater harvesting encourages a greater connection to and appreciation for the natural world. By actively engaging in the process of collecting and using rainwater, we become more aware of our environment and the impact our actions have on it. This heightened sense of responsibility can inspire us to adopt other sustainable practices, ultimately contributing to a greener and more resilient future for our planet.

In conclusion, rainwater harvesting is an essential component of sustainable living. By embracing this practice, we can conserve water, protect our ecosystems, save money, promote social equity, and foster a deeper connection with the environment. As we move forward in our quest for a more sustainable world, rainwater harvesting will undoubtedly play a pivotal role in shaping our collective future.

Understanding the Basic Components of a Rainwater Harvesting System

Rainwater harvesting is a simple yet effective method of collecting and storing rainwater for later use. It is an environmentally friendly practice that promotes sustainable living by reducing our reliance on municipal water supplies. This section will explore the fundamental components of a rainwater harvesting system and how they work together to provide a reliable water source for your home.

A typical rainwater harvesting system consists of the following key components:

Catchment Surface: The catchment surface is where rainwater is collected, usually on your home's roof. The size and material of your roof will determine the amount of water you can collect. Generally, larger roofs made of smooth, non-porous materials like metal or tiles are more effective at capturing rainwater.

Gutters and Downspouts: These are essential for directing the rainwater from your roof to the storage tank. Gutters should be properly maintained and cleaned regularly to prevent debris from clogging the system and reducing efficiency. Downspouts should be designed to minimize splashing and overflow, ensuring the maximum amount of water is collected.

Leaf Screens and First Flush Diverters: Leaf screens are installed at the entrance of the downspouts to prevent leaves and other debris from entering the system. First, flush diverters separate the initial dirty water from the rest of the rainwater, ensuring that only clean water enters the storage tank. This helps maintain the stored water's quality and reduces the need for treatment.

Storage Tank: The storage tank is the heart of the rainwater harvesting system. It is where the collected rainwater is stored for later use. Storage tanks can be made from various materials, such as plastic, concrete, or metal, and come in different sizes to suit your needs. The tank size will depend on the water you wish to store and the space available on your property.

Filtration and Treatment: Depending on the intended use of the harvested rainwater, you may need to install a filtration and treatment system—this range from simple sediment filters to more advanced UV or chemical treatment systems. Filtration and treatment are essential if you use the harvested water for drinking or other domestic purposes.

Delivery System: The delivery system distributes the stored rainwater to various points of use within your home. Depending on your needs and preferences, this can include a simple gravity-fed system or a more complex pump-driven system. The delivery system should be designed to ensure that the water pressure and flow rates are suitable for your intended uses.

Overflow and Drainage: It is essential to have a proper overflow and drainage system in place to prevent your storage tank from overflowing during heavy rainfall. This can be achieved by installing an overflow pipe that directs excess water away from the tank and into a safe drainage area.

By understanding the essential components of a rainwater harvesting system, you can make informed decisions when designing and installing your system. The next section will discuss assessing your home's rainwater harvesting potential and determining the most suitable method for your needs.

Assessing Your Home's Rainwater Harvesting Potential

Rainwater harvesting is an excellent way to reduce water consumption, save money on water bills, and contribute to a more sustainable lifestyle. However, before you dive into the world of rainwater harvesting, it's essential to assess your home's potential for collecting and utilizing rainwater effectively. In this section, we will guide you through evaluating your home's rainwater harvesting potential, considering factors such as roof size, rainfall patterns, and water usage.

Roof Size and Material

The first step in assessing your home's rainwater harvesting potential is to determine the size and material of your roof. The roof area is crucial in determining the amount of rainwater you can collect, as a larger surface area will collect more water. To calculate the size of your roof, measure the length and width of your home and multiply these figures together.

The material of your roof also plays a significant role in the efficiency of your rainwater harvesting system. Ideally, your roof should be made of a non-toxic material that does not leach harmful chemicals into the collected water. Some suitable materials include metal, slate, and clay tiles. If your roof is made of less desirable material, such as asphalt shingles, consider installing a pre-filtration system to ensure the collected water is safe.

Rainfall Patterns

The next factor to consider is the average annual rainfall in your area. This information can usually be found through local weather stations or online resources. Knowing the average rainfall will help you estimate the amount of water you can collect and determine the storage tank size needed to accommodate your needs.

Remember that rainfall patterns can vary significantly from year to year, so planning for both average and extreme conditions is essential. In addition, consider the seasonal distribution of rainfall, as this can affect the efficiency of your system and the need for additional water storage during drier months.

Water Usage

To maximize the benefits of your rainwater harvesting system, it's essential to have a clear understanding of your household's water usage. Begin by analyzing your water bills to determine your average monthly consumption. Then, consider how to use rainwater in your daily activities, such as watering plants, flushing toilets, and even washing clothes. By identifying the areas where rainwater can replace or supplement your current water usage, you can better estimate the storage tank size and filtration system needed to meet your needs.

Site Assessment

Finally, conduct a thorough site assessment of your property to identify potential challenges or limitations for installing a rainwater harvesting system. Some factors to consider include the slope of your land, the location of your home, your property boundaries, and the presence of any underground utilities or obstacles that may interfere with the installation of your system.

By assessing your home's rainwater harvesting potential, you can ensure that your investment in a rainwater harvesting system is well-planned and tailored to your specific needs. In the next section, we will discuss the legal and regulatory considerations for rainwater harvesting, helping you navigate the process of obtaining the necessary permits and approvals for your system.

Legal and Regulatory Considerations for Rainwater Harvesting

As we delve deeper into the world of rainwater harvesting, it's essential to address the legal and regulatory aspects that govern this practice. While rainwater harvesting is a sustainable and eco-friendly solution to water scarcity, it is crucial to ensure that your system adheres to the laws and regulations in your area. In this section, we will discuss the legal landscape surrounding rainwater harvesting and provide guidance on navigating the regulatory framework to ensure your system is compliant.

Understanding the Legal Landscape

The legal landscape for rainwater harvesting varies significantly from one jurisdiction to another. In some regions, rainwater harvesting is encouraged through incentives and rebates. In others, it may be subject to strict regulations or even prohibited. This diversity in regulations stems from differences in water rights, local climate conditions, and the perceived impact of rainwater harvesting on the environment and existing water infrastructure.

To ensure that your rainwater harvesting system is legally compliant, familiarize yourself with your area's laws and regulations. This may involve researching local building codes, zoning ordinances, and water rights laws and consulting with local authorities and experts in the field.

Navigating the Permitting Process

In many jurisdictions, installing a rainwater harvesting system requires obtaining permits from local authorities. The permitting process typically involves submitting detailed plans of your proposed system, including information on the system's components, capacity, and intended use. Additionally, you may be required to demonstrate that your system meets specific safety, health, and environmental standards.

To navigate the permitting process smoothly, consulting with professionals experienced in rainwater harvesting system design and installation is advisable. These experts can help you develop a compliant system and guide you through the necessary steps to obtain the required permits.

Compliance with Building Codes and Zoning Ordinances

Your rainwater harvesting system must comply with local building codes and zoning ordinances. These regulations often dictate the allowable size, location, and appearance of rainwater harvesting systems and the materials and construction methods used.

For example, some building codes may require that rainwater storage tanks be constructed of specific materials or meet certain structural standards. Similarly, zoning ordinances may restrict the placement of rainwater harvesting systems in certain areas or require that they be screened from view.

To ensure compliance with these regulations, consult with local building and zoning officials and seek professional assistance in designing and installing your system.

Water Rights Considerations

In some jurisdictions, water rights laws may impact your ability to harvest rainwater. These laws govern the ownership and use of water resources and, in some cases, may restrict the collection and use of rainwater.

Suppose you live in an area with strict water rights laws. In that case, it is crucial to research these regulations and consult with legal experts to determine if rainwater harvesting is permissible and under what conditions.

Staying Informed and Adapting to Changes

As the importance of sustainable water management continues to gain recognition, the legal and regulatory landscape for rainwater harvesting is likely to evolve. To ensure ongoing compliance, stay informed about changes in laws and regulations in your area and be prepared to adapt your system as needed.

In conclusion, understanding and navigating the legal and regulatory considerations for rainwater harvesting is crucial to implementing a successful and compliant system. By researching local laws, consulting with experts, and staying informed about regulation changes, you can confidently embrace rainwater harvesting as a sustainable solution for your home and contribute to a greener future.

Embracing Rainwater Harvesting for a Greener Future

As we reach the end of this introduction, it is essential to take a moment to reflect on the significance of this sustainable practice in our daily lives. By now, you have been introduced to the importance of rainwater harvesting for sustainable living, the essential components of a rainwater harvesting system, how to assess your home's potential for rainwater harvesting, and the legal and regulatory considerations involved in implementing such a system. With this knowledge, you are ready to continue to the next chapter.

Embracing rainwater harvesting is not just about conserving water and reducing your environmental footprint. It is also about adopting a mindset that values our planet's resources and seeks to preserve them for future generations. By implementing a rainwater harvesting system in your home, you are taking a significant step towards sustainable living and contributing to the global effort to combat water scarcity and climate change.

Moreover, rainwater harvesting offers numerous benefits beyond environmental sustainability. Utilizing this natural resource can reduce your reliance on municipal water supplies, lower your water bills, and even improve the quality of the water you use for various purposes around your home. Additionally, rainwater harvesting systems can be customized to suit your specific needs and preferences, allowing you to create a solution that works best for your household.

As you embark on your rainwater harvesting journey, remember that every drop counts. Even small-scale systems can significantly impact conserving water and promoting a more sustainable lifestyle. Furthermore, sharing your experiences and knowledge with others can inspire your community to adopt similar practices and create a ripple effect of positive change.

In conclusion, rainwater harvesting is a practical, efficient, and environmentally responsible solution to address growing water scarcity and climate change concerns. By embracing and incorporating this practice into your home, you can play a crucial role in safeguarding our planet's precious resources and fostering a greener future for all. So, continue reading and take the plunge – the rewards of rainwater harvesting await you!

1

THE IMPORTANCE OF RAINWATER HARVESTING FOR YOUR HOME

A serene forest glen in the midst of a gentle rain, with water droplets glistening on the leaves.

In today's rapidly changing world, where the demand for natural resources is ever-increasing, adopting sustainable practices that can help preserve our environment and ensure a healthy future for future generations is crucial. One such practice that has gained significant attention in recent years is **rainwater harvesting.** As the name suggests, rainwater harvesting is collecting, storing, and utilizing rainwater for various purposes, such as irrigation, gardening, and even domestic use after proper treatment. This section will explore the importance of rainwater harvesting for your home and how it can contribute to a more sustainable and eco-friendly lifestyle.

Rainwater harvesting is not new; it has been practiced for centuries across different civilizations. However, with the advent of modern plumbing systems and the easy availability of water, this ancient practice was largely forgotten. But as the world faces increasing water scarcity and environmental challenges, rainwater harvesting has once again emerged as a viable and essential solution.

The importance of rainwater harvesting for your home cannot be overstated. From environmental benefits to financial advantages, rainwater harvesting offers a multitude of reasons for homeowners to consider implementing this practice. In the following sections, we will delve deeper into these benefits, discussing how rainwater harvesting can improve water security, enhance the health of your landscape and garden, and contribute to a more sustainable future for all.

By the end of this chapter, you will have a comprehensive understanding of the significance of rainwater harvesting for your home. You will be inspired to take action toward a greener and more sustainable lifestyle. So, let's embark on this journey together and discover the incredible potential of rainwater harvesting in transforming our lives and the world around us.

Environmental Benefits of Rainwater Harvesting

Rainwater harvesting is a practical solution for homeowners looking to reduce their water bills and increase their self-sufficiency. It also offers a range of environmental benefits that contribute to a healthier planet. In this section, we will explore how rainwater harvesting can positively impact the environment and help preserve our precious water resources.

First and foremost, rainwater harvesting **reduces the demand for our existing water supply systems.** By collecting and using rainwater, homeowners can significantly decrease their reliance on municipal water sources, which are often overburdened and subject to shortages during periods of drought. This reduced demand helps conserve water resources and ensures enough water is available for essential uses such as drinking, sanitation, and agriculture.

Another environmental benefit of rainwater harvesting is the **reduction of stormwater runoff.** When rain falls on impervious surfaces such as roofs, driveways, and streets, it can quickly accumulate and cause flooding, erosion, and water pollution. By capturing rainwater before it becomes runoff, homeowners can help mitigate these issues and protect local waterways from contamination. This is particularly important in urban areas, where stormwater runoff can carry pollutants such as oil, pesticides, and heavy metals into rivers and streams, harming aquatic life and degrading water quality.

Rainwater harvesting also contributes to **energy conservation.** Treating and distributing water requires a significant amount of energy. By using rainwater instead of relying solely on treated water, homeowners can help reduce the energy consumption associated with water provision. This, in turn, helps to decrease greenhouse gas emissions and our overall carbon footprint.

Furthermore, rainwater is a naturally soft water source, containing fewer minerals and chemicals than treated water. This can benefit your garden and landscape, **reducing the risk of chemical buildup** in the soil and promoting healthier plant growth. Additionally, using rainwater for irrigation can help prevent the overuse of chemical fertilizers and pesticides, which can harm the environment and pollute our waterways.

In conclusion, the environmental benefits of rainwater harvesting are numerous and far-reaching. By embracing this sustainable practice, homeowners can play an active role in conserving water resources, protecting local ecosystems, and promoting a healthier environment for future generations. Furthermore, as we continue to face the challenges of climate change and water scarcity, rainwater harvesting offers a practical and eco-friendly solution that can significantly impact our collective efforts to preserve the planet.

Financial Advantages of Collecting Rainwater

In today's world, where the cost of living is constantly rising, finding ways to save money and reduce expenses is essential. One often overlooked method of cutting costs is rainwater harvesting. By collecting and utilizing rainwater, homeowners can enjoy significant financial advantages. In this section, we will explore how rainwater harvesting can lead to monetary savings and increased financial stability.

First and foremost, rainwater harvesting **reduces your reliance on municipal water supplies.** By capturing and storing rainwater, you can decrease your consumption of treated water from the city or town. This reduction in usage directly translates to lower water bills, leaving more money in your pocket. Additionally, as water rates continue to rise, the savings from rainwater harvesting will only become more substantial.

Another financial advantage of rainwater harvesting is the potential for reduced stormwater management costs. In many urban areas, stormwater fees are levied on property owners based on the impervious surface of their property. By implementing rainwater harvesting systems, you can reduce the volume of stormwater runoff from your property, lowering your stormwater fees.

Moreover, rainwater harvesting can lead to savings in energy costs. Municipal water treatment and distribution require a significant amount of energy, which is often reflected in the cost of water. By using rainwater for non-potable purposes, such as flushing toilets and irrigation, you can reduce the demand for treated water and, in turn, the energy required to produce it. This can lead to lower energy bills and a smaller carbon footprint.

Furthermore, rainwater harvesting can contribute to increased property value. A well-designed and maintained rainwater harvesting system can attract potential homebuyers, particularly those who are environmentally conscious or interested in self-sufficiency. By investing in a rainwater harvesting system, you can increase the resale value of your home, making it a wise financial decision in the long run.

Finally, government incentives and rebates are often available for homeowners who invest in rainwater harvesting systems. These incentives can offset the initial costs of installing a system, making it a more affordable and financially viable option: local research programs and rebates to take full advantage of these opportunities.

In conclusion, the financial advantages of collecting rainwater are numerous and wide-ranging. From reduced water and energy bills to increased property value and government incentives, rainwater harvesting is a wise investment for homeowners looking to save money and contribute to a more sustainable future.

Improving Water Security and Self-Sufficiency

Water security and self-sufficiency are becoming increasingly important for homeowners in today's rapidly changing world. As climate change continues to disrupt traditional water sources and urbanization puts pressure on existing infrastructure, it is crucial to explore alternative methods to ensure a consistent and reliable water supply for our homes. Rainwater harvesting is one such method that can significantly contribute to improving water security and self-sufficiency.

Water security refers to the availability of an adequate quantity and quality of water for various household needs, such as drinking, cooking, and sanitation. By collecting and storing rainwater, homeowners can reduce their dependence on municipal water supplies, which may be subject to shortages or contamination. This is particularly important in regions where water scarcity is a pressing issue, as it helps to alleviate the burden on already strained resources.

Moreover, rainwater harvesting systems can serve as a **backup water source during emergencies or disruptions** in the public water supply. In a natural disaster or infrastructure failure, having a reserve of stored rainwater can provide a vital lifeline for families, ensuring they have access to clean water for their essential needs.

Self-sufficiency, on the other hand, refers to the ability of a household to meet its water requirements independently without relying on external sources. By harvesting rainwater, homeowners can significantly reduce their water bills and gain greater control over their water usage. This leads to financial savings and fosters a sense of empowerment and self-reliance as individuals take responsibility for their water management.

Additionally, rainwater harvesting systems can be tailored to suit each household's specific needs and preferences. Homeowners can choose from various collection methods, storage capacities, and treatment options, allowing them to design a system that best meets their unique requirements. This flexibility further enhances the self-sufficiency aspect of rainwater harvesting, enabling individuals to adapt their systems to changing circumstances or evolving needs.

In conclusion, rainwater harvesting offers a practical and sustainable solution for improving water security and self-sufficiency in our homes. By embracing this method, homeowners can safeguard their water supply and contribute to a more resilient and resource-efficient future.

Enhancing Landscape and Garden Health

One of the most noticeable and rewarding advantages of rainwater harvesting for your home is its positive impact on the health and vitality of your landscape and garden. By utilizing the natural resource that falls from the sky, you can provide your plants with the nourishment they need to thrive while promoting a sustainable and eco-friendly approach to gardening.

Rainwater is an excellent source of hydration for your plants compared to tap water. This is because rainwater is free of chemicals, such as chlorine and fluoride, commonly found in municipal water supplies. These chemicals can harm the delicate balance of your garden's ecosystem, potentially causing damage to your plants and soil.

In contrast, rainwater is naturally soft and contains essential nutrients and minerals beneficial for plant growth.

When you harvest rainwater for your garden, you are not only providing your plants with a healthier water source but also helping to reduce the demand for our overburdened water supply systems. This is particularly important in areas where water scarcity is a growing concern. In addition, by using rainwater to irrigate your landscape, you can contribute to conserving this precious resource and reducing the strain on our environment.

Another advantage of rainwater harvesting is that it can help **improve the overall soil quality** in your garden. As rainwater percolates through the soil, it helps to break up compacted earth, allowing for better aeration and drainage. In addition, this process encourages the growth of beneficial microorganisms and earthworms, which are essential for maintaining a healthy soil structure. In turn, this leads to stronger root systems and more robust plants.

In addition to the benefits for your plants and soil, rainwater harvesting can contribute to a more attractive and diverse landscape. By designing your rainwater collection system to include features such as rain gardens, swales, or permeable paving, you can create an aesthetically pleasing environment supporting various plant species and wildlife. This enhances the beauty of your outdoor space and promotes biodiversity and ecological balance.

In conclusion, incorporating rainwater harvesting into your home's landscape and garden design can bring numerous benefits to the health and vitality of your plants, soil, and overall ecosystem. By embracing this sustainable practice, you can contribute to conserving our planet's resources while enjoying a thriving and beautiful outdoor sanctuary.

Embracing Rainwater Harvesting for a Sustainable Future

In conclusion, rainwater harvesting is an essential practice that benefits homeowners and the environment. By embracing this sustainable approach to water management, we can contribute to a greener and more resilient future for our communities and the planet.

Throughout this chapter, we have explored the environmental benefits of rainwater harvesting, such as reducing the strain on our water resources, minimizing stormwater runoff, and promoting groundwater recharge. These advantages help preserve our ecosystems and contribute to a healthier environment for future generations.

Financially, rainwater harvesting systems can significantly save water bills, making it an attractive investment for homeowners. As water scarcity and the cost of water continue to rise, these savings will become increasingly valuable. Moreover, government incentives and rebates can further sweeten the deal, making rainwater harvesting an even more appealing option.

Water security and self-sufficiency are also vital aspects of rainwater harvesting. By collecting and storing rainwater, homeowners can reduce their dependence on municipal water supplies and ensure a reliable water source during drought or water restrictions. In addition, this increased self-reliance can provide peace of mind and a sense of empowerment, knowing that you are taking control of your water needs.

Furthermore, rainwater harvesting can enhance the health and beauty of your landscape and garden. The natural, chemical-free quality of rainwater is ideal for nurturing plants, leading to more vibrant and flourishing green spaces around your home. This not only improves the aesthetics of your property but also contributes to the overall well-being of the environment.

In summary, rainwater harvesting is a practical and sustainable solution that can benefit your home and the environment. By adopting this eco-friendly practice, you can actively conserve our precious water resources, save money, and promote a greener future for all. So, let us embrace rainwater harvesting and positively impact our world, one drop at a time.

Chapter Summary

1. Rainwater harvesting is a sustainable practice that involves collecting, storing, and utilizing rainwater for various purposes, such as irrigation, gardening, and domestic use after proper treatment.
2. Environmental benefits of rainwater harvesting include reducing the demand on existing water supply systems, minimizing stormwater runoff, conserving energy, and providing a naturally soft water source for gardens and landscapes.
3. Financial advantages of collecting rainwater include lower water bills, reduced stormwater management costs, potential energy savings, increased property value, and government incentives and rebates.
4. Rainwater harvesting can improve water security and self-sufficiency by reducing dependence on municipal water supplies and serving as a backup water source during emergencies or disruptions.
5. Utilizing rainwater for landscape and garden irrigation can enhance plant health and soil quality. It is free of chemicals found in treated water and contains essential nutrients and minerals.
6. Rainwater harvesting systems can be tailored to suit each household's specific needs and preferences, allowing for flexibility and adaptability.
7. Incorporating rainwater harvesting into landscape design can create a more attractive and diverse environment, supporting various plant species and wildlife.
8. Embracing rainwater harvesting contributes to a greener and more resilient future for communities and the planet, conserving water resources, promoting biodiversity, and fostering a more sustainable lifestyle.

2

UNDERSTANDING THE RAINWATER HARVESTING PROCESS

A rainwater tank made entirely out of recycled materials, sitting in the middle of a lush garden.

Rainwater harvesting is an ancient practice utilized by various civilizations throughout history. Today, it continues to gain popularity as a sustainable and eco-friendly solution to water scarcity and the increasing demand for clean water. This chapter will delve into the fascinating world of rainwater harvesting, exploring its importance, components, and the steps required to implement a successful system for your home.

At its core, rainwater harvesting is collecting, storing, and utilizing rainwater for various purposes, such as irrigation, flushing toilets, and even potable water when adequately treated. This simple yet effective method allows homeowners to reduce their reliance on municipal water supplies and decrease their water bills while contributing to a greener and more sustainable environment.

The beauty of rainwater harvesting lies in its **adaptability**. Whether you live in an urban apartment or a sprawling suburban home, a rainwater harvesting system can be tailored to suit your needs. The possibilities are endless, from small-scale systems that collect water in barrels to more complex setups that integrate with your home's plumbing.

In the following sections, we will discuss the importance of rainwater harvesting for your home and the benefits it can provide. We will also examine the various components and design elements of a rainwater harvesting system, guiding you through the steps necessary to implement your system. Finally, we will cover the essential maintenance and troubleshooting aspects to ensure that your rainwater harvesting system remains efficient and effective for years.

So, let's embark on this exciting journey towards sustainable water management and discover how rainwater harvesting can transform your home and your relationship with this precious resource.

The Importance of Rainwater Harvesting for Your Home

Rainwater harvesting is not just an eco-friendly practice but a vital step towards sustainable living and responsible water management. Yet, as homeowners, we often take our water supply for granted, not realizing our consumption's impact on the environment and our wallets. This section will delve into the importance of rainwater harvesting for your home, exploring its numerous benefits and how it contributes to a greener, more sustainable future.

First and foremost, rainwater harvesting helps **conserve a precious resource: water**. With climate change and increasing urbanization, water scarcity is becoming a pressing issue worldwide. Collecting and utilizing rainwater directly reduces your reliance on municipal water supplies, often sourced from overburdened rivers, lakes, and aquifers. This, in turn, helps preserve these water sources for future generations and supports the overall health of our ecosystems.

Another significant advantage of rainwater harvesting is the reduction of your water bills. As you rely more on harvested rainwater for household needs, such as gardening, flushing toilets, and even laundry, you will notice a considerable decrease in your monthly water expenses. This financial benefit appeals to homeowners looking to cut costs and invest in long-term, sustainable solutions.

Rainwater harvesting also reduces stormwater runoff, which can cause soil erosion, flooding, and pollution in our waterways. By capturing rainwater on your property, you are minimizing the volume of water that flows into storm drains and sewer systems. This helps protect your home from potential water damage and benefits the environment by reducing the pollutants that enter our rivers, lakes, and oceans.

Furthermore, rainwater is naturally soft and chemical-free, making it an ideal choice for gardening and irrigation. Plants thrive on rainwater, as it is free of the salts, minerals, and other impurities commonly found in tap water. This results in healthier, more vibrant gardens and landscapes, which can also increase your property's aesthetic appeal and value.

Lastly, rainwater harvesting promotes a sense of environmental stewardship and self-sufficiency. By implementing a rainwater collection system, you are actively managing your water resources and reducing your ecological footprint. This benefits the planet and instills a sense of pride and accomplishment in your efforts to live a more sustainable lifestyle.

In conclusion, rainwater harvesting is essential for homeowners who wish to conserve water, save money, and contribute to a healthier environment. By understanding the importance of rainwater harvesting and incorporating it into your daily life, you are taking a significant step towards sustainable water management and a greener future.

Components and Design of a Rainwater Harvesting System

Designing an efficient rainwater harvesting system for your home is exciting and rewarding. Not only will it help you conserve water and reduce your water bills, but it will also contribute to a more sustainable environment. This section will discuss the essential components of a rainwater harvesting system and how to design one that suits your needs and preferences.

A well-designed rainwater harvesting system consists of several key components, each playing a crucial role in collecting, storing, and distributing rainwater. Let's take a closer look at these components:

Catchment Surface: The catchment surface is where rainwater is collected, such as on rooftops or other hard surfaces. The size and material of your catchment surface will directly impact the amount of water you can harvest. Therefore, you should choose a surface that is large, clean, and made of non-toxic materials.

Gutters and Downspouts: These components channel rainwater from the catchment surface to the storage tank. They should be adequately sized and installed to ensure efficient water flow and prevent clogging. Regular cleaning and maintenance of gutters and downspouts are essential to keep your system functioning optimally.

First Flush Diverter: This device helps to divert the initial flow of rainwater, which may contain contaminants and debris from the catchment surface. Doing so ensures that only clean water enters the storage tank, reducing the need for extended filtration and maintenance.

Storage Tank: The storage tank is where the collected rainwater is stored for later use. The size of your tank will depend on your water needs, available space, and budget. Common materials for storage tanks include plastic, concrete, and metal. Choosing a tank that is durable, UV-resistant, and well-insulated to prevent algae growth and maintain water quality is essential.

Filtration System: Before using the harvested rainwater, it's crucial to filter it to remove any remaining impurities. Various filtration methods are available, including sediment filters, activated carbon filters, and UV sterilization. The filtration system choice will depend on the intended use of the harvested water and the quality of the water collected.

Distribution System: Once the rainwater is filtered, it must be distributed throughout your home for various uses, such as irrigation, toilet flushing, or drinking. This can be achieved through a network of pipes, pumps, and valves that connect the storage tank to the points of use.

When designing your rainwater harvesting system, it's essential to consider factors such as local climate, rainfall patterns, and water usage habits. This will help you determine your system's appropriate size and capacity, ensuring that it meets your water needs while remaining cost-effective and environmentally friendly.

In conclusion, a well-designed rainwater harvesting system is an invaluable addition to any home. By understanding the various components and design considerations, you can create a system that saves you money and contributes to a more sustainable future.

Steps to Implementing a Rainwater Harvesting System

Now that you have a solid understanding of the rainwater harvesting process and its components, it's time to dive into the steps involved in implementing a rainwater harvesting system for your home. By following these steps, you'll be well on your way to enjoying the benefits of sustainable water management and reducing your reliance on traditional water sources.

Assess your water needs: Before you begin, it's essential to determine your household's water usage and requirements. Consider factors such as the size of your family, the types of plants in your garden, and your daily water consumption habits. This information will help you design a system that meets your specific needs.

Analyze rainfall patterns: Study the rainfall patterns in your area to determine the average annual rainfall and the distribution of rain throughout the year. This will help you estimate the amount of water you can harvest and guide you in selecting the appropriate size of your storage tank.

Choose a suitable location: Select a location for your rainwater harvesting system close to the areas where you plan to use the water, such as your garden or laundry room. Ensure the site has adequate space for the storage tank and is easily accessible for maintenance.

Design your system: With your water needs and rainfall patterns in mind, design a rainwater harvesting system that includes components such as gutters, downspouts, filters, storage tanks, and a distribution system. Consult a professional to ensure your system is designed efficiently and safely.

Obtain permits and approvals: Depending on your local regulations, you may need permits or approvals before installing your rainwater harvesting system. Check with your local authorities to ensure you comply with any necessary regulations.

Install your system: Once your design is complete and you have obtained the necessary permits, you can begin the installation process. This may involve hiring a professional installer or tackling the project yourself if you have the necessary skills and tools. Be sure to follow all safety guidelines and manufacturer recommendations during installation.

Test and adjust: After installing your system, ensure it functions correctly. This may involve checking for leaks, ensuring the filters are working properly, and verifying that the water is being distributed as intended. Then, make any necessary adjustments to optimize the performance of your system.

Educate your household: Teach your family members about the rainwater harvesting system and how to use the harvested water responsibly. Please encourage them to adopt water-saving habits to maximize the benefits of your new system.

By following these steps, you'll be well on your way to implementing a rainwater harvesting system that meets your household's needs and contributes to sustainable water management. In the next section, we'll discuss the maintenance and troubleshooting of your rainwater harvesting system to ensure its longevity and continued performance.

Maintenance and Troubleshooting of Your Rainwater Harvesting System

As with any home system, proper maintenance and troubleshooting are essential to ensure the longevity and efficiency of your rainwater harvesting system. In this section, we will delve into the necessary steps to keep your system running smoothly and address any potential issues that may arise.

Regular maintenance of your rainwater harvesting system is crucial for its optimal performance. By following a consistent maintenance schedule, you can prevent potential problems, ensure the quality of the harvested water, and prolong the life of your system components. Here are some essential maintenance tasks to keep in mind:

Inspect and clean gutters and downspouts: Regularly check your gutters and downspouts for debris, such as leaves and twigs, which can obstruct the flow of rainwater. Cleaning them at least twice a year, or more frequently if you live in an area with heavy foliage, will help maintain the efficiency of your system.

Check the roof and catchment surface: Periodically inspect your roof and catchment surface for any damage or wear that could affect the quality of the harvested rainwater. Repair any issues promptly to prevent contamination or leakage.

Monitor the pre-filtration system: Examine your pre-filtration systems for clogs or damage, such as leaf screens and first flush diverters. Clean and replace components to ensure your system effectively filters out debris and contaminants.

Inspect storage tanks and connections: Regularly check your storage tanks for leaks, cracks, or signs of wear. Additionally, examine all connections and seals to ensure they remain watertight. Repair or replace any damaged components as necessary.

Test water quality: Periodically test the quality of your harvested rainwater to ensure it meets the necessary standards for its intended use. If you notice any changes in water quality, investigate the cause and take corrective action as needed.

Maintain pumps and treatment systems: If your rainwater harvesting system includes pumps or treatment systems, follow the manufacturer's recommended maintenance schedule to ensure proper functioning.

Despite diligent maintenance, issues may still arise with your rainwater harvesting system. Troubleshooting these problems promptly can help prevent further damage and ensure the continued efficiency of your system. Here are some common issues and their potential solutions:

Reduced water flow: If you notice a decrease in the flow of harvested rainwater, check for clogs or obstructions in your gutters, downspouts, and pre-filtration system. Additionally, inspect your storage tank for sediment buildup, which can impede water flow.

Water quality issues: If your harvested rainwater has an unpleasant odor, taste, or appearance, investigate the source of the contamination. This may include damaged roofing materials, pollution from nearby sources, or inadequate filtration. Then, address the issue accordingly to restore water quality.

Leaks or overflow: Inspect the tank and connections for damage if you observe leaks or overflow from your storage tank. Additionally, ensure that your tank is adequately filled and your overflow system functions correctly.

By following a regular maintenance schedule and addressing any issues promptly, you can ensure the longevity and efficiency of your rainwater harvesting system. In doing so, you are embracing sustainable water management at home and contributing to a greener and more resilient future for our planet.

Embracing Sustainable Water Management at Home

In conclusion, rainwater harvesting is an essential and sustainable approach to water management that every homeowner should consider implementing. By embracing this eco-friendly practice, you contribute to conserving our precious water resources, reducing your dependence on municipal water supplies, lowering your water bills, and promoting a healthier environment for future generations.

Throughout this chapter, we have explored the importance of rainwater harvesting for your home, delved into the components and design of an effective rainwater harvesting system, and discussed the steps to implementing and maintaining such a system. With this knowledge, you can make informed decisions about incorporating rainwater harvesting into your home's water management strategy.

As you embark on this journey towards sustainable water management, remember that every drop of rainwater you collect and utilize is a step towards a greener and more self-sufficient home. By adopting rainwater harvesting, you benefit your household, set a positive example for your community, and inspire others to follow suit.

In a world where water scarcity is becoming increasingly pressing, we all must take responsibility for our water usage and strive to make a difference. Rainwater harvesting is a practical and effective way to do just that. By implementing this system in your home, you play a vital role in preserving our planet's most precious resource.

So, embrace sustainable water management at home – for your wallet, community, and the environment. With every raindrop you collect, you make a lasting impact and contribute to a more sustainable future.

Chapter Summary

1. Rainwater harvesting is an ancient, sustainable, and eco-friendly practice that helps conserve water, reduce reliance on municipal water supplies, and decrease water bills.

2. The importance of rainwater harvesting lies in its contribution to preserving water resources, reducing water bills, minimizing stormwater runoff, promoting healthier gardens, and fostering environmental stewardship.
3. A well-designed rainwater harvesting system consists of several key components, including catchment surface, gutters and downspouts, first flush diverter, storage tank, filtration system, and distribution system.
4. To implement a rainwater harvesting system, assess your water needs, analyze rainfall patterns, choose a suitable location, design your system, obtain permits and approvals, install the system, test and adjust, and educate your household.
5. Regular maintenance of your rainwater harvesting system is crucial for optimal performance, including inspecting and cleaning gutters and downspouts, checking the roof and catchment surface, monitoring the pre-filtration system, inspecting storage tanks and connections, testing water quality, and maintaining pumps and treatment systems.
6. Troubleshooting common issues, such as reduced water flow, water quality issues, and leaks or overflow can help prevent further damage and ensure the continued efficiency of your system.
7. By following a regular maintenance schedule and addressing any issues promptly, you can ensure the longevity and efficiency of your rainwater harvesting system, contributing to a greener and more resilient future for our planet.
8. Embracing sustainable water management at home through rainwater harvesting benefits your household, sets a positive example for your community, and inspires others to adopt eco-friendly practices.

3

ASSESSING YOUR HOME'S RAINWATER HARVESTING POTENTIAL

A futuristic, sleek roof and gutter system that uses advanced technology to automatically detect and redirect rainwater.

Imagine a world where you could reduce your water bill, contribute to a sustainable environment, and have a reliable source of water for your home, all at the same time. Sounds too good to be true? Well, rainwater harvesting is a practical and eco-friendly solution that can make this dream a reality. But before you jump into this exciting venture, assessing your home's rainwater harvesting potential is ***essential.*** This chapter will guide you through evaluating your home's suitability for rainwater harvesting and determining the feasibility of implementing such a system.

Rainwater harvesting is collecting, storing, and using rainwater for various purposes, such as irrigation, flushing toilets, and even drinking, after proper treatment. This ancient practice has gained renewed interest in recent years due to increasing water scarcity, rising water costs, and a growing awareness of the need for sustainable living. However, not all homes are equally suited for rainwater harvesting, and it's crucial to determine whether your home has the necessary conditions to make this endeavor worthwhile.

In this chapter, we will explore various factors that contribute to your home's rainwater harvesting potential. We will begin by evaluating your roof and gutter system, which are crucial in capturing and directing rainwater to your storage system. Next, we will analyze your home's water usage and needs to determine the potential benefits of rainwater harvesting. Finally, we will also delve into local climate and rainfall patterns, as they significantly impact the amount of rainwater you can collect.

Legal and regulatory factors are another essential aspect to consider, as some regions have specific laws and regulations regarding rainwater harvesting. Finally, we will wrap up the chapter with a conclusion to help you determine the feasibility of rainwater harvesting for your home.

By the end of this chapter, you will clearly understand your home's rainwater harvesting potential and the necessary steps to take if you decide to pursue this sustainable and cost-effective solution. So, let's embark on this exciting journey and discover the possibilities rainwater harvesting can offer for your home and the environment.

Evaluating Your Roof and Gutter System

Before diving into the world of rainwater harvesting, it's essential to evaluate the current state of your roof and gutter system. This section will guide you through assessing your home's existing infrastructure to determine if it's suitable for rainwater collection and what modifications may be necessary.

First, let's discuss the importance of having a well-maintained roof and gutter system. Your roof is the primary catchment area for rainwater. At the same time, the gutters and downspouts direct the water to your storage tanks. A poorly maintained or damaged roof can lead to water contamination, leaks, and structural issues, which can negatively impact the efficiency and safety of your rainwater harvesting system.

To begin your evaluation, inspect the material and condition of your roof. The ideal roofing material for rainwater harvesting is metal, as it's durable, easy to clean, and less likely to leach contaminants into the water. Other suitable materials include tile, slate, and asphalt shingles. However, suppose your roof is made of wood, clay, or asbestos. In that case, it's not recommended for rainwater collection due to the potential for contamination or structural issues.

Next, examine the overall condition of your roof. Look for signs of damage, such as missing or broken shingles, rust, or visible leaks. If you find any issues, **addressing them before installing a rainwater harvesting system is crucial**. Additionally, ensure that your roof is free of debris, such as leaves, branches, or moss, as these can obstruct water flow and lead to contamination.

Now, let's turn our attention to the gutter system. Gutters and downspouts are vital in directing rainwater from your roof to your storage tanks. To assess their suitability for rainwater harvesting, check for proper installation, adequate size, and good condition. Gutters should be securely fastened to your home without visible gaps or leaks. Downspouts should be large enough to handle the volume of water your roof collects during a rainstorm and should direct water away from your home's foundation.

It's also essential to keep your gutters clean and free of debris. Clogged gutters can cause water to overflow, leading to water damage and reducing the amount of rainwater you can collect. Regularly cleaning your gutters and installing gutter guards can help prevent these issues.

Lastly, consider the layout of your gutter system and the location of your downspouts. Downspouts should be positioned near the intended location of your storage tanks to minimize the distance water must travel and reduce the potential for contamination. You can modify your gutter system or add downspouts to optimize your rainwater collection if necessary.

In conclusion, evaluating your roof and gutter system is crucial in determining the feasibility of rainwater harvesting for your home. Ensure your roof is made of suitable material, in good condition, and free of debris. By maintaining a well-functioning gutter system, you'll be well on your way to reaping the benefits of this sustainable and cost-effective water source.

Analyzing Your Home's Water Usage and Needs

Before diving into the world of rainwater harvesting, it's essential to have a clear understanding of your home's water usage and needs. This will help you determine the appropriate size and design of your rainwater harvesting system and ensure that it meets your household's requirements effectively. This section will discuss various aspects of analyzing your home's water usage and needs, which will ultimately help you make informed decisions about your rainwater harvesting project.

To begin with, start by examining your **water bills** for the past year. This will give you a rough estimate of your household's monthly water consumption. Remember that water usage may vary seasonally, with higher consumption during the warmer months due to gardening, lawn maintenance, and filling swimming pools. Please note these fluctuations, as they will play a crucial role in determining the capacity of your rainwater harvesting system.

Next, consider the specific water needs of your household. List all the activities and appliances that require water, such as drinking, cooking, bathing, laundry, and irrigation. It's important to remember that not all harvested rainwater needs to be treated to potable standards. For instance, water used for irrigation or toilet flushing can be lower quality than water used for drinking or cooking. By identifying the various uses of water in your home, you can design a rainwater harvesting system that caters to these specific needs, ultimately saving you money on water treatment costs.

Now that you clearly understand your home's water usage, it's time to evaluate the potential of rainwater harvesting to meet these needs. Start by calculating the percentage of your household's water usage that can be realistically replaced with harvested rainwater. This will depend on factors such as the size of your roof, the efficiency of your gutter system, and local rainfall patterns. Remember that replacing 100% of your water needs with rainwater may not be feasible, especially in areas with low rainfall or strict regulations regarding rainwater use.

Finally, consider the financial aspect of implementing a rainwater harvesting system. While the initial investment may seem high, weighing this against the long-term savings on your water bills is essential. Additionally, local incentives or rebates may be available for homeowners who invest in rainwater harvesting systems, further reducing the overall cost.

In conclusion, analyzing your home's water usage and needs is critical in determining the feasibility of rainwater harvesting for your household. By carefully considering your water consumption patterns, specific water needs, and the potential for rainwater harvesting to meet these needs, you can make informed decisions about the size and design of your rainwater harvesting system. This will help you maximize the benefits of this sustainable and environmentally-friendly practice while reducing your reliance on municipal water supplies and lowering your water bills.

Identifying Local Climate and Rainfall Patterns

Understanding your area's local climate and rainfall patterns is a crucial step in assessing your home's rainwater harvesting potential. By gaining insight into the amount and frequency of rainfall, you can better determine the size and type of rainwater harvesting system that will be most effective for your home. This section will discuss the importance of analyzing climate data, how to access reliable information, and how to use this knowledge to optimize your rainwater harvesting system.

First, let's discuss why it's essential to understand your local climate and rainfall patterns. The amount of rainwater you can collect depends on the frequency and intensity of rainfall in your area. A minor rainwater harvesting system may suffice in regions with abundant rainfall. In contrast, a more extensive system may be necessary to meet your water needs in areas with less frequent or lower-intensity rainfall. Additionally, knowing the seasonal variations in rainfall can help you plan for periods of water scarcity or surplus.

You can consult various sources such as local meteorological departments, government agencies, or online weather databases to access accurate and up-to-date climate data. These resources typically provide historical rainfall data, which can give you a good idea of your area's average annual and monthly precipitation levels. Therefore, it's essential to use reliable sources to ensure that your rainwater harvesting system is designed based on accurate information.

Once you have gathered the necessary climate data, you can **analyze the information** to determine the best rainwater harvesting system for your home. Start by calculating the average monthly rainfall and identifying any seasonal patterns. For example, suppose you live in an area with a distinct rainy season. In that case, you may need to invest in a larger storage tank to collect and store water during the drier months. On the other hand, if rainfall is evenly distributed throughout the year, a smaller tank may be sufficient.

It's also important to consider the intensity of rainfall events. For example, suppose your area experiences heavy downpours. In that case, your rainwater harvesting system should be designed to handle the increased flow rate, which may require more oversized gutters and downspouts. Conversely, a system with smaller components may be adequate if your region has more frequent, lighter rainfall.

In conclusion, identifying local climate and rainfall patterns is critical to assessing your home's rainwater harvesting potential. By understanding the amount, frequency, and intensity of rainfall in your area, you can design a system tailored to your specific needs and optimizes the collection and storage of this valuable resource. Remember to consult reliable sources for climate data and consider seasonal variations when planning your rainwater harvesting system.

Considering Legal and Regulatory Factors

As you embark on the journey to harness the power of rainwater for your home, it is crucial to consider the legal and regulatory factors that may impact your rainwater harvesting project. While rainwater harvesting is generally encouraged as an environmentally friendly and sustainable practice, specific rules and regulations within your local area may govern its implementation. This section will discuss the importance of understanding these legal and regulatory factors and guide how to navigate them.

First and foremost, it is essential to familiarize yourself with any local, state, or federal laws and regulations that pertain to rainwater harvesting. These laws can vary widely depending on your location, so it is essential to **research and understand** the specific requirements for your area. For example, some regions may have strict guidelines on the size and type of rainwater collection systems allowed. In contrast, others might offer incentives for implementing such systems. In some cases, permits may be required to install rainwater harvesting systems, especially if they involve significant alterations to your home's structure or plumbing.

In addition to legal requirements, you must consider any regulations or guidelines set forth by your homeowners' association (HOA) or other local governing bodies. These organizations may have specific rules regarding the appearance and placement of rainwater collection systems, mainly if they are visible from the street or neighboring properties. Review relevant documents, such as your HOA's bylaws or covenants, conditions, and restrictions (CC&Rs), to ensure your planned rainwater harvesting system complies with these rules.

Another critical aspect to consider is the potential impact of your rainwater harvesting system on your neighbors and the surrounding environment. While rainwater collection can provide numerous benefits, it is essential to ensure that your system does not cause any harm or inconvenience to others. For example, ensure your system does not divert water from neighboring properties or create standing water that could attract pests or create safety hazards. If you need clarification on your system's potential impacts, consult a professional.

Finally, staying informed about any changes to legal and regulatory requirements that may affect your rainwater harvesting system is essential. Laws and regulations can evolve, and staying up-to-date on these changes can help you maintain compliance and avoid any potential issues. Consider subscribing to newsletters or joining local organizations focused on sustainability and water conservation to stay informed about relevant updates.

In conclusion, understanding and adhering to the legal and regulatory factors surrounding rainwater harvesting is a crucial step in ensuring the success of your project. By taking the time to research and comply with these requirements, you can confidently move forward with your rainwater harvesting journey, knowing that your system is both environmentally responsible and legally sound.

Determining the Feasibility of Rainwater Harvesting for Your Home

As we reach the end of this chapter, it's time to take a step back and consider the bigger picture. By now, you should understand the various factors contributing to your home's rainwater harvesting potential. From evaluating your roof and gutter system to analyzing your home's water usage and needs, you've gathered valuable information that will help you decide whether rainwater harvesting is a viable option for your household.

In addition to the practical aspects of rainwater harvesting, it's essential to consider your area's local **climate and rainfall patterns.** After all, the success of your rainwater harvesting system will largely depend on the amount and frequency of rainfall your region experiences. By identifying these patterns, you can better estimate the potential benefits of implementing a rainwater harvesting system and adjust your design as needed.

Of course, knowing any legal and regulatory factors that may impact your ability to harvest rainwater is crucial. For example, while many regions encourage rainwater harvesting as a sustainable practice, others may have restrictions or guidelines that you'll need to follow. By familiarizing yourself with these regulations, you can ensure your system is compliant and avoid potential issues.

So, is rainwater harvesting feasible for your home? Ultimately, the answer will depend on your unique circumstances and the information you've gathered throughout this chapter. Suppose, for example, suppose you've discovered that your roof and gutter system are well-suited for rainwater collection. In that case, your household water usage aligns with the potential benefits of a rainwater harvesting system. Furthermore, if your local climate and regulations are favorable, rainwater harvesting could be a valuable addition to your home.

However, if you've identified significant challenges or limitations in these areas, weighing the potential benefits against the costs and effort required to overcome these obstacles is essential. In some cases, exploring alternative water-saving strategies or modifying your home to improve its rainwater harvesting potential may be more practical.

Regardless of your final decision, assessing your home's rainwater harvesting potential has undoubtedly provided you with valuable insights into your household's water usage and the environmental factors at play. By taking the time to understand these aspects, you're better equipped to make informed choices about your home's water management and contribute to a more sustainable future.

Chapter Summary

1. Rainwater harvesting is a practical and eco-friendly solution that can reduce water bills, contribute to a sustainable environment, and provide a reliable water source for your home.
2. Evaluating your roof and gutter system is crucial for determining the feasibility of rainwater harvesting, as they play a significant role in capturing and directing rainwater to your storage system.
3. Analyzing your home's water usage and needs helps determine the appropriate size and design of your rainwater harvesting system and ensures it effectively meets your household's requirements.
4. Understanding local climate and rainfall patterns is essential for designing a rainwater harvesting system tailored to your specific needs and optimizing the collection and storage of rainwater.
5. Familiarizing yourself with legal and regulatory factors surrounding rainwater harvesting is essential to ensure your system is compliant.
6. Considering the potential impact of your rainwater harvesting system on neighbors and the surrounding environment is essential for responsible and effective implementation.
7. Staying informed about legal and regulatory requirements changes can help you maintain compliance and adapt your rainwater harvesting system as needed.
8. Determining the feasibility of rainwater harvesting for your home depends on your unique circumstances, including your roof and gutter system, household water usage, local climate, and regulations.

4

DESIGNING AND PLANNING YOUR RAINWATER HARVESTING SYSTEM

A green, eco-friendly house with a rainwater system installation that collects rainwater from the roof.

Rainwater harvesting is an ancient practice that has renewed interest recently as homeowners seek sustainable and eco-friendly solutions to reduce their water consumption and reliance on municipal water supplies. A well-designed rainwater harvesting system can provide a **valuable source of water** for your home and contribute to the conservation of water resources and the reduction of stormwater runoff. This chapter will guide you through the essential steps of designing and planning a rainwater harvesting system tailored to your home's unique needs and characteristics.

Designing and planning a rainwater harvesting system may seem daunting. Still, it can be an enjoyable and rewarding experience with the right approach and guidance. The key to a successful rainwater harvesting system is careful planning, considering your home's specific requirements and the local climate conditions. By following the steps outlined in this chapter, you will be able to create a system that is efficient and effective in meeting your water needs while also being aesthetically pleasing and easy to maintain.

In the following sections, we will explore the various aspects of designing and planning a rainwater harvesting system, including assessing your home's rainwater collection potential, choosing the right components, creating a maintenance and monitoring plan, and integrating the system with your home's existing infrastructure. By the end of this chapter, you will clearly understand the essential elements of a rainwater harvesting system and be well-equipped to bring your vision to life.

So, let's embark on this exciting journey towards a more sustainable and water-efficient home by diving into the world of rainwater harvesting system design and planning.

Assessing Your Home's Rainwater Collection Potential

Before diving into the rainwater harvesting world, evaluating your home's rainwater collection potential is essential. This assessment will help you determine the feasibility of installing a rainwater harvesting system and guide you in selecting the right components to maximize your water savings. This section will walk you through the key factors to consider when assessing your home's rainwater collection potential.

Roof Area: The size of your roof plays a significant role in determining the amount of rainwater you can collect. Generally, the larger the roof area, the more rainwater you can harvest. To calculate your roof's catchment area, measure the length and width of your roof and multiply them together. Remember that roofs with steeper slopes may have a larger catchment area than flat roofs.

Rainfall Patterns: The amount of rainfall in your region is another crucial factor. Research the average annual rainfall in your area and note any seasonal variations. This information will help you estimate the volume of rainwater you can collect and decide on the appropriate storage capacity for your system.

Roof Material: The type of roofing material on your home can affect the quality and quantity of rainwater you can collect. Metal roofs, for example, are ideal for rainwater harvesting as they are smooth and allow for efficient water flow. On the other hand, roofs made of asphalt shingles or clay tiles may require additional filtration systems to ensure the collected water is clean and safe for use.

Gutter and Downspout System: Your home's gutter and downspout system is crucial in directing rainwater from your roof to your storage tank. Inspect your gutters and downspouts for any signs of damage or blockages that may hinder water flow. Additionally, consider the size and layout of your gutter system to ensure it can handle the rainwater you plan to collect.

Available Space for Storage: The space for installing a storage tank is another critical consideration. Assess your property to identify suitable locations for a rainwater storage tank, considering factors such as accessibility for maintenance, distance from your home's foundation, and any local regulations or zoning requirements.

Intended Use of Collected Rainwater: Consider how you plan to use the harvested rainwater. Suppose you intend to use it for non-potable purposes, such as irrigation or toilet flushing. In that case, you may not require extensive filtration or treatment systems. However, suppose you plan to use the collected rainwater for drinking or cooking. In that case, you must invest in a more advanced purification system to ensure the water is safe for consumption.

By carefully assessing your home's rainwater collection potential, you can make informed decisions about the design and components of your rainwater harvesting system. This evaluation will help you maximize your water savings and ensure that your system is tailored to your specific needs and circumstances.

Choosing the Right Rainwater Harvesting Components

Selecting the appropriate components for your rainwater harvesting system is crucial to its efficiency and longevity. You can create a practical and aesthetically pleasing system by understanding the various options available and how they fit into your overall design. This section will discuss the critical components of a rainwater harvesting system and guide how to choose the best options for your home.

Catchment Surface: The catchment surface is where rainwater is collected, typically on your home's roof. The material and slope of your roof will impact the quality and quantity of water collected. Metal roofs are ideal for rainwater harvesting, as they are durable, low-maintenance, and provide a smooth surface for water to flow. Asphalt shingles, while common, may leach chemicals into the water and require additional filtration. Consider your roof's material and condition when planning your system.

Gutters and Downspouts: Gutters and downspouts are essential for directing water from your roof to your storage tank. Choose durable, corrosion-resistant materials, and easy to clean, such as aluminum or stainless steel. Ensure your gutters are appropriately sized to handle the water your roof collects. In addition, install leaf guards to prevent debris from entering your system.

Pre-filtration: Pre-filtration is essential in maintaining water quality and preventing debris from entering your storage tank. Various pre-filtration options include leaf screens, first-flush diverters, and roof washers. Leaf screens prevent large debris from entering your downspouts. At the same time, first flush diverters and roof washers remove the initial dirty water from your roof, ensuring only clean water enters your storage tank.

Storage Tank: The storage tank is the heart of your rainwater harvesting system, and choosing the right size and material is crucial. Your tank should be large enough to store water during heavy rainfall and provide a sufficient supply during dry spells. Consider your home's water usage, local rainfall patterns, and available space when selecting a tank size. Tanks can be made from various materials, including polyethylene, fiberglass, and stainless steel. Each material has pros and cons, so research and compare options before deciding.

Pump and Pressure System: To utilize the harvested rainwater throughout your home, you will need a pump and pressure system. Select an energy-efficient and powerful pump to meet your home's water pressure requirements. Additionally, consider installing a pressure tank to maintain consistent water pressure and reduce the frequency of pump cycling.

Filtration and Disinfection: To ensure the harvested rainwater is safe, it must be adequately filtered and disinfected. Filtration options include sediment filters, activated carbon filters, and ultraviolet (UV) filters, which remove particles, chemicals, and pathogens. Depending on your water quality and intended use, you may need a combination of these filters. For disinfection, UV light or chlorination systems can kill any remaining bacteria or viruses.

Overflow and Drainage: Finally, your rainwater harvesting system should have an overflow mechanism to prevent your storage tank from overfilling during heavy rainfall. This overflowing water can be directed to a garden, stormwater drain, or another appropriate location.

By carefully selecting the right components for your rainwater harvesting system, you can create a reliable, efficient, and environmentally friendly water source for your home. Consider your home's unique needs and local regulations when choosing components, and consult with professionals if necessary. With proper planning and execution, your rainwater harvesting system will be a valuable addition to your home for years.

Creating a Maintenance and Monitoring Plan

A well-designed rainwater harvesting system is a valuable addition to your home. Still, like any other infrastructure, it requires regular maintenance and monitoring to ensure optimal performance and longevity. In this section, we will discuss the importance of creating a comprehensive maintenance and monitoring plan and the key steps to include in your plan to keep your rainwater harvesting system running smoothly.

Regular maintenance of your rainwater harvesting system is essential for several reasons:

1. It helps to prevent the buildup of debris, algae, and other contaminants that can affect the quality of the harvested rainwater.
2. It ensures that all system components, such as gutters, filters, and storage tanks, function correctly and efficiently.
3. A well-maintained system can reduce the risk of leaks, water damage, and other issues that can lead to costly repairs or even compromise the structural integrity of your home.

To create an effective maintenance and monitoring plan, consider the following steps:

Schedule regular inspections: Set a schedule for inspecting your rainwater harvesting system, including gutters, downspouts, filters, storage tanks, and any other components. This can be done monthly, quarterly, or annually, depending on your system's complexity and the local climate. During these inspections, check for signs of wear and tear, leaks, or damage, and make any necessary repairs or replacements.

Clean and maintain gutters and downspouts: Regularly clean your gutters and downspouts to prevent the buildup of leaves, dirt, and other debris that can impede the flow of rainwater and lead to clogs or damage. This is especially important during the fall and spring seasons, when debris accumulation is more likely.

Monitor and clean filters: Check your rainwater filters regularly to ensure they are not clogged or damaged. Clean or replace filters as needed to maintain the quality of the harvested water and prevent strain on your system's components.

Inspect and maintain storage tanks: Regularly inspect your storage tanks for signs of leaks, corrosion, or other damage. Clean the interior of the tanks periodically to prevent the growth of algae and bacteria and ensure that the tank's vents and overflow systems are functioning correctly.

Monitor water quality: Regularly test the quality of your harvested rainwater to ensure it meets the appropriate standards for its intended use, such as irrigation, toilet flushing, or even potable water if adequately treated. This can help you identify issues with your system's filtration or treatment processes and make necessary adjustments.

Keep a maintenance log: Document all maintenance and monitoring activities, including dates, findings, and any repairs or replacements. This log can serve as a valuable reference for future maintenance. In addition, it can help you identify patterns or recurring issues requiring further attention.

Stay informed about local regulations and best practices: Regularly review local regulations and guidelines for rainwater harvesting to ensure your system remains compliant. Additionally, stay informed about new technologies, best practices, and other advancements in rainwater harvesting to optimize your system's performance and efficiency.

By following these steps and creating a comprehensive maintenance and monitoring plan, you can ensure the longevity and effectiveness of your rainwater harvesting system. This will help you conserve water and reduce your environmental footprint and save money on water bills and potential repair costs in the long run.

Integrating Your Rainwater Harvesting System with Your Home's Existing Infrastructure

As you embark on the exciting journey of designing and planning your rainwater harvesting system, it is essential to consider how it will seamlessly integrate with your home's existing infrastructure. This integration not only ensures the efficiency and effectiveness of your rainwater harvesting system but also maintains your home's aesthetic appeal and functionality. This section will discuss various aspects of integrating your rainwater harvesting system with your home's existing infrastructure, including plumbing, electrical, and landscaping considerations.

First and foremost, evaluating your **home's plumbing system** is crucial. If adequately treated, your rainwater harvesting system must connect to your existing plumbing to supply water for various household uses, such as irrigation, toilet flushing, and even potable water. Consult a professional plumber to determine the best approach for connecting your rainwater harvesting system to your home's plumbing. This may involve installing additional pipes, rerouting existing pipes, or integrating a pump system to ensure adequate water pressure.

Next, consider the **electrical requirements** of your rainwater harvesting system. Some components, such as pumps and control systems, may require electrical connections. Working with a licensed electrician is essential to ensure that your system is safely and correctly wired, adhering to all local codes and regulations. Additionally, you plan to incorporate a solar-powered pump or other renewable energy sources. In that case, you must consider the appropriate electrical connections and equipment for these components.

Landscaping is another critical aspect of integrating your rainwater harvesting system with your home's infrastructure. Thoughtful landscaping can enhance the visual appeal of your rainwater harvesting system and improve its efficiency. For instance, strategically placed plants can help direct rainwater toward your catchment area. At the same time, porous paving materials can facilitate water infiltration into the ground, reducing runoff. Furthermore, consider incorporating rain gardens or swales into your landscape design to help manage excess rainwater and prevent flooding.

Lastly, ensuring that your rainwater harvesting system complies with all local building codes, regulations, and permitting requirements is essential. This may involve obtaining permits, working with a licensed contractor, or submitting plans for review by local authorities. Compliance with these regulations ensures the safety and effectiveness of your rainwater harvesting system and protects your investment in your home.

In conclusion, integrating your rainwater harvesting system with your home's existing infrastructure is crucial in designing and planning your system. By carefully considering plumbing, electrical, and landscaping aspects, as well as adhering to local codes and regulations, you can create a rainwater harvesting system that is both functional and aesthetically pleasing. With thoughtful integration, your rainwater harvesting system will become a valuable addition to your home, providing a sustainable water source for years.

Bringing Your Rainwater Harvesting System to Life

In conclusion, designing and planning a rainwater harvesting system for your home is exciting and rewarding. By following the steps outlined in this chapter, you can create a sustainable and efficient system that benefits the environment and reduces your water bills and reliance on municipal water supplies.

As you bring your rainwater harvesting system to life, remember that the key to success lies in carefully planning, selecting the right components, and integrating the system seamlessly with your home's infrastructure. Doing so ensures that your system operates at peak performance and provides you with a **reliable water source for years.**

Remember the importance of creating a maintenance and monitoring plan to keep your system in top shape. Regular maintenance will help you avoid potential issues and ensure that your rainwater harvesting system remains valuable to your home.

Lastly, remember that rainwater harvesting is about more than just its practical benefits. It is also an opportunity to contribute to a more sustainable future and become a responsible steward of our planet's precious resources. By harvesting rainwater, you are taking a significant step towards conserving water, reducing your carbon footprint, and promoting environmental awareness within your community.

So, as you embark on this journey to bring your rainwater harvesting system to life, take pride in knowing that you are making a positive impact on the world around you. With careful planning, dedication, and a little creativity, you can create a functional and environmentally friendly system, providing you and your family with a sustainable source of water for years to come.

Chapter Summary

1. Rainwater harvesting is an ancient practice that has gained renewed interest due to its sustainability and eco-friendliness, helping homeowners reduce water consumption and reliance on municipal water supplies.
2. Assessing your home's rainwater collection potential is crucial, considering factors such as roof area, rainfall patterns, roof material, gutter and downspout systems, available space for storage, and intended use of collected rainwater.
3. Choosing the right components for your rainwater harvesting system is essential for its efficiency and longevity, including catchment surfaces, gutters and downspouts, pre-filtration, storage tanks, pumps, pressure systems, filtration and disinfection, and overflow and drainage systems.
4. Creating a comprehensive maintenance and monitoring plan is vital for ensuring optimal performance and longevity of your rainwater harvesting system, including regular inspections, cleaning and maintaining gutters and downspouts, monitoring and cleaning filters, inspecting and maintaining storage tanks, monitoring water quality, and staying informed about local regulations and best practices.
5. Integrating your rainwater harvesting system with your home's infrastructure is crucial for seamless functionality and aesthetic appeal, considering plumbing, electrical, and landscaping aspects.

6. Compliance with local building codes, regulations, and permitting requirements is essential for the safety and effectiveness of your rainwater harvesting system and for protecting your investment in your home.
7. Thoughtful landscaping can enhance the visual appeal of your rainwater harvesting system and improve its efficiency, incorporating elements such as strategically placed plants, permeable paving materials, rain gardens, or swales.
8. Designing and planning a rainwater harvesting system is an exciting and rewarding endeavor that contributes to a more sustainable future, conserves water, reduces carbon footprint, and promotes environmental awareness within your community.

5

CHOOSING THE RIGHT RAINWATER COLLECTION AND STORAGE EQUIPMENT

A green, eco-friendly house with a rainwater system installation that collects rainwater from the roof.

Rainwater harvesting has been vital for centuries, offering numerous benefits such as reducing water bills, conserving precious resources, and promoting a sustainable lifestyle. With the growing awareness of environmental issues and the need to conserve water, many homeowners now consider rainwater harvesting an essential part of their home management. This chapter will explore the various rainwater collection and storage equipment available and guide you in making the right choices for your home's rainwater harvesting system.

Rainwater collection and storage equipment are **essential** for any rainwater harvesting system. With proper treatment, these systems are designed to capture, store, and distribute rainwater for various household uses, such as watering plants, flushing toilets, and even drinking. By investing in the right equipment, you can maximize the efficiency of your rainwater harvesting system and enjoy the numerous benefits it offers.

The process of rainwater harvesting begins with the collection of rainwater from surfaces such as roofs, gutters, and downspouts. This water is then channeled through a series of filters and storage tanks before being distributed for use. The choice of equipment for each stage of this process plays a crucial role in determining your rainwater harvesting system's overall efficiency and effectiveness.

This chapter will discuss the various types of rainwater collection systems, storage solutions, and critical factors to consider when choosing the right equipment for your home. We will also provide practical tips and insights to help you make informed decisions and ensure your rainwater harvesting system is tailored to your needs and preferences.

By the end of this chapter, you will have a comprehensive understanding of the different rainwater collection and storage equipment available and be well-equipped to make the right choices for your home's rainwater harvesting system. So, let's dive in and explore the fascinating world of rainwater harvesting equipment!

Assessing Your Home's Rainwater Harvesting Potential

Before diving into the world of rainwater harvesting, evaluating your home's potential for collecting and utilizing rainwater effectively is essential. This assessment will understand your home's unique characteristics and help you make informed decisions when selecting the proper collection and storage equipment. This section will discuss the key factors to consider when assessing your home's rainwater harvesting potential.

Roof Size and Material

The first step in assessing your home's rainwater harvesting potential is to evaluate the size and material of your roof. The roof serves as the primary catchment area for rainwater, and its size directly impacts the volume of water you can collect. Measure the square footage of your roof to estimate the amount of rainwater you can harvest.

Additionally, consider the material of your roof, as this will affect the quality of the collected rainwater. For example, metal roofs are ideal for rainwater harvesting, as they are less likely to contaminate water. On the other hand, roofs made of asphalt shingles or treated wood may release chemicals that could compromise the quality of the harvested rainwater. Finally, suppose your roof is made of something other than a suitable material. In that case, consider alternative catchment surfaces or invest in a pre-filtration system.

Rainfall Patterns and Climate

Your region's rainfall will significantly impact your home's rainwater harvesting potential. First, research the average annual rainfall in your area and note any seasonal patterns. This information will help you determine the size and capacity of your rainwater collection and storage system. A smaller storage tank may suffice in areas with consistent rainfall throughout the year. However, a larger storage capacity will be necessary to ensure an adequate water supply during dry periods in regions with distinct wet and dry seasons.

Water Usage and Demand

Consider your household's water usage and demand to design a rainwater harvesting system effectively. Analyze your water bills to determine your average monthly water consumption. Next, identify the activities and appliances that use the most water, such as gardening, laundry, and toilet flushing. This information will help you estimate the percentage of your water needs that can be met through rainwater harvesting and guide you in selecting the appropriate equipment.

Local Regulations and Restrictions

Before implementing a rainwater harvesting system, it's crucial to familiarize yourself with any local regulations and restrictions that may apply. For example, some regions may have specific guidelines on the types of collection systems allowed or require permits for installing storage tanks. Additionally, there may be restrictions on harvesting rainwater, particularly for potable purposes. Consult your local authorities and adhere to regulations to ensure a safe and compliant rainwater harvesting system.

By carefully assessing your home's rainwater harvesting potential, you can make informed decisions when selecting the proper collection and storage equipment. The following section will explore the different types of rainwater collection systems and their respective advantages and disadvantages.

Comparing Different Types of Rainwater Collection Systems

Rainwater harvesting is an environmentally friendly and cost-effective way to utilize a natural resource that often goes to waste. However, not all rainwater collection systems are created equal. This section will explore the various types of rainwater collection systems available, their advantages and disadvantages, and how to determine which one is best suited for your home.

Roof Catchment Systems

The most common type of rainwater collection system is the roof catchment system. As the name suggests, this system collects rainwater from your home's roof. The water is channeled through gutters and downspouts, then directed into a storage tank or cistern. Roof catchment systems are popular because they are relatively easy to install and can be adapted to fit most homes.

Advantages:

- Easy to install and maintain
- It can be adapted to most homes
- Efficient in collecting large volumes of water

Disadvantages:

- Water quality may be affected by roofing materials and debris
- Requires regular cleaning and maintenance of gutters and downspouts

Surface Catchment Systems

Surface catchment systems collect rainwater from ground-level surfaces, such as driveways, patios, and other paved areas. These systems typically use channels, grates, and pipes to direct water into a storage tank or cistern. Surface catchment systems are ideal for homes with limited roof space or those looking to maximize their rainwater harvesting potential.

Advantages:

- Can collect water from multiple sources
- Ideal for homes with limited roof space

Disadvantages:

- It may require more extensive installation and maintenance
- Water quality may be affected by surface contaminants

Rain Gardens and Bioswales

Rain gardens and bioswales are landscape features that capture and filter rainwater runoff. These systems use plants and soil to slow water flow, allowing it to infiltrate the ground and recharge groundwater supplies. While less efficient at collecting water for immediate use, rain gardens and bioswales are excellent options for homeowners looking to improve their property's overall water management and reduce stormwater runoff.

Advantages:

- Improves overall water management on your property
- Reduces stormwater runoff and associated pollution
- Enhances the aesthetic appeal of your landscape

Disadvantages:

- Not as efficient for immediate water collection and use
- Requires careful planning and maintenance to ensure proper function

When comparing these different types of rainwater collection systems, it's essential to consider your home's unique characteristics, such as roof size, available space, and water usage needs. By carefully evaluating your options and weighing the advantages and disadvantages of each system, you can make an informed decision that will help you maximize your rainwater harvesting potential and contribute to a more sustainable future.

Selecting the Ideal Storage Solutions for Your Needs

Now that you better understand the various rainwater collection systems available, it's time to delve into the crucial storage aspect. Choosing the proper storage solution for your harvested rainwater is essential to ensure an efficient and effective system is in place. In this section, we will explore the different storage options available, discuss their pros and cons, and provide guidance on selecting the ideal solution for your needs.

Types of Rainwater Storage Solutions

Several types of storage solutions are available for harvested rainwater, each with its own advantages and disadvantages. Some of the most common options include:

Above-ground tanks: These are the most common and affordable type of storage solution. They are easy to install and maintain and can be made from plastic, metal, or concrete. However, they can be less aesthetically pleasing and may take up valuable yard space.

Underground tanks are installed underground, making them less visible and more space-efficient. They are also less susceptible to temperature fluctuations, which can help maintain water quality. However, they can be more expensive to install and may require additional maintenance.

Rain barrels are small, above-ground containers that collect rainwater from downspouts. They are affordable and straightforward for those with limited space or budget. However, they have a smaller storage capacity and may require more frequent emptying.

Modular storage systems: These systems consist of multiple small tanks connected, allowing for storage capacity and location flexibility. They can be installed above or below ground and quickly expanded as needed. However, they may be more expensive than single-tank solutions.

Factors to Consider When Choosing a Storage Solution

When selecting the ideal storage solution for your rainwater harvesting system, consider the following factors:

Storage capacity: Determine the amount of rainwater you expect to collect and use based on your home's harvesting potential and water usage needs. Choose a storage solution that can accommodate this volume while allowing for extra capacity in case of heavy rainfall or increased usage.

Space availability: Consider the available space in your property for installing a storage solution. Above-ground tanks and rain barrels may require more yard space, while underground tanks and modular systems can be more space-efficient.

Aesthetics: If the appearance of your storage solution is essential to you, consider options that blend well with your home's exterior or can be easily concealed, such as underground tanks or modular systems.

Budget: The cost of storage solutions can vary significantly, so consider your budget when deciding. While above-ground tanks and rain barrels are generally more affordable, investing in a more expensive option like an underground tank may be worthwhile in the long run due to its durability and lower maintenance requirements.

Maintenance: Different storage solutions may require varying maintenance levels, such as cleaning, repairs, or replacement. Consider the long-term maintenance requirements of each option and choose one that best suits your lifestyle and preferences.

Selecting the ideal storage solution for your rainwater harvesting system is crucial in ensuring its efficiency and effectiveness. By considering factors such as storage capacity, space availability, aesthetics, budget, and maintenance requirements, you can make an informed decision that best suits your needs and preferences. With the proper storage solution, you can enjoy the numerous benefits of rainwater harvesting, from conserving water resources to reducing utility bills and promoting a more sustainable lifestyle.

Key Factors to Consider When Choosing Equipment

When selecting the right equipment for your home's rainwater harvesting system, there are several key factors to consider. By considering these factors, you can ensure that you choose the most suitable and efficient equipment for your needs. This section will discuss the most important aspects to consider when selecting your rainwater collection and storage equipment.

Quality and Durability: One of the most important factors to consider when choosing equipment for your rainwater harvesting system is the quality and durability of the materials used. Investing in high-quality equipment made from durable materials will ensure that your system remains functional and practical for many years. In addition, look for products made from corrosion-resistant materials, such as stainless steel, UV-resistant plastics, or fiberglass, as these will be less likely to degrade.

Capacity: Another crucial factor to consider when selecting equipment is the capacity of the collection and storage systems. The capacity of your rainwater harvesting system should be based on your household's water usage and the average rainfall in your area. It's essential to choose equipment that can handle the volume of water you expect to collect and provide adequate storage for times when rainfall is scarce.

Ease of Installation and Maintenance: When choosing equipment for your rainwater harvesting system, it's essential to consider how easy it will be to install and maintain. Look for products with clear installation instructions that require minimal tools and expertise to set up. Additionally, consider the maintenance requirements of the equipment, such as cleaning and replacing filters, and ensure that you or a professional can efficiently perform these tasks.

Cost and Budget: Your budget will undoubtedly play a significant role in the equipment you choose for your rainwater harvesting system. While investing in high-quality, durable equipment is essential, it's also important to find products that fit your budget. Be sure to compare the prices and features of different products to find the best value for your money.

Aesthetics: While the primary purpose of your rainwater harvesting system is to collect and store water, it's also essential to consider the visual impact of the equipment on your property. Choose equipment that complements the design and style of your home, and consider options that can be easily concealed or integrated into your landscape.

Local Regulations and Requirements: Before selecting equipment for your rainwater harvesting system, research any local regulations and requirements that may apply. Some areas may have specific guidelines regarding the types of materials that can be used, the size of storage tanks, or the installation process. Ensure the equipment you choose complies with these regulations to avoid potential issues.

In conclusion, taking the time to carefully consider these key factors when choosing equipment for your home's rainwater harvesting system will help ensure you make the best choices for your needs. By selecting high-quality, durable equipment that meets your capacity requirements, is easy to install and maintain, fits within your budget, and complies with local regulations, you can create an efficient and effective rainwater harvesting system that will serve your household for years to come.

Making the Right Choices for Your Home's Rainwater Harvesting System

In conclusion, rainwater harvesting is environmentally friendly and cost-effective for meeting your home's water needs. Choosing the proper collection and storage equipment ensures that your rainwater harvesting system operates efficiently and effectively. This chapter has guided you to **make informed decisions** when selecting the appropriate equipment for your home's rainwater harvesting system.

To recap, assessing your home's rainwater harvesting potential is essential by considering factors such as your local climate, roof size, and water usage requirements. This information will help you determine the best rainwater collection system for your needs, whether a simple gutter and downspout system or a more advanced system with filters and pumps.

Regarding storage solutions, you must consider capacity, material, and location factors. Selecting the right storage tank or cistern ensures that your harvested rainwater remains clean and safe. Additionally, it is crucial to consider the key factors when choosing equipment, such as durability, ease of maintenance, and cost.

As you embark on your rainwater harvesting journey, remember that the choices you make regarding equipment will significantly impact your system's overall success and efficiency. By taking the time to research and select the best equipment for your needs, you can enjoy the numerous benefits that come with rainwater harvesting, including reduced water bills, a more sustainable lifestyle, and the satisfaction of contributing to a greener planet.

Ultimately, the right choices for your home's rainwater harvesting system will depend on your unique circumstances and preferences. By carefully considering your options and following the guidance in this chapter, you can create a customized rainwater harvesting system that meets your needs and helps you achieve your sustainability goals. Happy harvesting!

Chapter Summary

1. Rainwater harvesting is an environmentally friendly and cost-effective solution for meeting your home's water needs, offering reduced water bills and a more sustainable lifestyle.
2. Assess your home's rainwater harvesting potential by considering roof size, material, climate, water usage, and regulations.
3. There are various rainwater collection systems, including roof catchment systems, surface catchment systems, and rain gardens or bioswales. Each has advantages and

disadvantages, so choose the one that best suits your needs.

4. Storage solutions for harvested rainwater include above-ground tanks, underground tanks, rain barrels, and modular storage systems. Consider storage capacity, space availability, aesthetics, budget, and maintenance requirements when choosing the right solution.
5. When selecting equipment for your rainwater harvesting system, prioritize quality and durability to ensure long-lasting functionality and effectiveness.
6. Consider the ease of installation and maintenance when choosing equipment, as this will impact your system's overall efficiency and user-friendliness.
7. Budget and local regulations are essential when selecting equipment for your rainwater harvesting system. Ensure your chosen equipment complies with local guidelines and fits within your budget.
8. By carefully considering your options and following the guidance in this chapter, you can create a customized rainwater harvesting system that meets your needs and helps you achieve your sustainability goals.

6

INSTALLING YOUR RAINWATER HARVESTING SYSTEM

A farmer collects water from a well in the middle of his lush green field.

This chapter will delve into the exciting process of installing your rainwater harvesting system. You have learned about the numerous benefits of rainwater harvesting, the various types of systems available, and the legal aspects to consider. Then, it is time to plunge and start the installation process.

Installing a rainwater harvesting system can be a rewarding and fulfilling experience. Not only will you be contributing to a more sustainable environment, but you will also be saving money on your water bills and ensuring a reliable water source for your home. This chapter will guide you through the entire process, from assessing your home and property for rainwater harvesting to choosing the right components and following a step-by-step installation guide.

The idea of installing a rainwater harvesting system might seem ***daunting*** at first. However, with the proper guidance and a little bit of patience, you will soon be on your way to enjoying the many benefits of this eco-friendly and cost-effective solution. Our goal is to provide you with clear and concise information and an engaging tone to make this process as enjoyable and stress-free as possible.

In the following sections, we will cover the essential steps to install your rainwater harvesting system successfully. We will provide a detailed, step-by-step guide to the installation process. By the end of this chapter, you will have all the knowledge and tools necessary to install your rainwater harvesting system confidently. So, let's get started on this exciting journey toward a more sustainable and self-sufficient home!

Step-by-Step Installation Process

Now that you have assessed your home, selected the right components for your rainwater harvesting system, and familiarized yourself with the necessary maintenance and troubleshooting tips, it's time to dive into the installation process. This section will guide you through a step-by-step process to ensure your rainwater harvesting system is installed correctly and efficiently.

Prepare the site: Clear where your rainwater harvesting system will be located before installation. Remove any debris, rocks, or vegetation that may obstruct the installation process or the system's functionality. Ensure the ground is level and stable to support the tank's weight when it's full of water.

Install the gutters and downspouts: If your home still needs them, you'll need to install them. Ensure they are appropriately secured and sloped to direct water toward the downspouts. Install leaf guards or gutter screens to prevent debris from entering the system.

Set up the catchment surface: The catchment surface is where rainwater is collected, usually on your roof. Ensure that your roof is clean and free of debris. Install a roof washer to filter out contaminants before the water enters your system.

Assemble and install the tank: Follow the manufacturer's instructions to assemble your storage tank. Once assembled, place the tank on a solid, level foundation near the downspout. Ensure that the tank is adequately secured and stable.

Connect the downspout to the tank: Attach a diverter or a flexible pipe to the downspout, directing the water flow into the storage tank. Make sure all connections are secure and watertight to prevent leaks.

Install the overflow system: An overflow system is essential to prevent your tank from overflowing during heavy rainfall. Connect an overflow pipe to the designated outlet on your tank and direct the excess water away from your home's foundation.

Set up the filtration system: Install a pre-tank filter to remove debris and contaminants before the water enters your storage tank. If your system includes a post-tank filter for additional purification, install it according to the manufacturer's instructions.

Install the pump and distribution system: If your rainwater harvesting system includes a pump, follow the manufacturer's guidelines for installation. Connect the pump to your storage tank and the distribution system, which may include hoses, irrigation systems, or indoor plumbing.

Test the system: Once everything is installed, test your rainwater harvesting system by running water through the gutters, downspouts, and storage tank. Check for any leaks or issues with the filtration, overflow, and distribution systems.

Make any necessary adjustments: After testing your system, make any necessary adjustments to ensure optimal performance. This may include tightening connections, adjusting the slope of your gutters, or fine-tuning the pump's settings.

By following these steps, you'll have a fully functional rainwater harvesting system that will provide you with a sustainable and eco-friendly water source for years. The following section will discuss maintenance and troubleshooting tips to keep your system running smoothly.

Enjoying the Benefits of Your Rainwater Harvesting System

Congratulations! You have successfully installed your rainwater harvesting system and are now ready to reap its numerous benefits. As you begin to utilize this eco-friendly and cost-effective solution, you will soon realize its positive impact on your home, wallet, and environment.

First and foremost, you will notice a **significant reduction in your water bills.** By harvesting rainwater, you are tapping into an accessible and renewable water source that can be used for various purposes around your home, such as watering your garden, flushing toilets, and even doing laundry. This not only saves you money but also helps conserve precious water resources.

In addition to the financial benefits, your rainwater harvesting system will contribute to a greener and more sustainable environment. By capturing and using rainwater, you reduce your reliance on municipal water supplies, often from overexploited sources like rivers and groundwater. This helps protect these vital resources and promotes a more balanced water cycle.

Moreover, your garden and plants will thrive with the use of rainwater. Unlike treated municipal water, rainwater is free of chemicals like chlorine and fluoride, which can harm your plants. Rainwater also contains natural minerals and nutrients that promote healthy plant growth, leading to a more vibrant and flourishing garden.

Another advantage of your rainwater harvesting system is its ability to reduce stormwater runoff. Capturing rainwater prevents it from flowing over paved surfaces and picking up pollutants like oil, pesticides, and fertilizers. This helps protect local waterways and ecosystems from contamination and reduces the risk of flooding in your community.

Lastly, your rainwater harvesting system will give you a sense of accomplishment and pride. By taking this step towards a more sustainable lifestyle, you are actively preserving our planet for future generations. Furthermore, your efforts will inspire others to follow suit and contribute to a global movement toward water conservation and environmental stewardship.

In conclusion, your rainwater harvesting system is a valuable addition to your home that offers numerous benefits. As you enjoy the financial savings, environmental advantages, and the satisfaction of contributing to a greener world, you can take pride knowing that you have made a wise and responsible choice. So, go ahead and enjoy the fruits of your labor, and let your rainwater harvesting system serve as a constant reminder of your commitment to a sustainable future.

Chapter Summary

1. Installing a rainwater harvesting system is a rewarding and fulfilling experience that contributes to a more sustainable environment, saves money on water bills, and ensures a reliable water source for your home.
2. The step-by-step installation process involves preparing the site, installing gutters and downspouts, setting up the catchment surface, assembling and installing the tank, connecting the downspout to the tank, installing the overflow system, setting up the filtration system, installing

the pump and distribution system, testing the system, and making any necessary adjustments.

3. The benefits of a rainwater harvesting system include reduced water bills, a greener and more sustainable environment, healthier plants and gardens, reduced stormwater runoff, and a sense of accomplishment and pride in contributing to a sustainable future.
4. By installing and maintaining a rainwater harvesting system, homeowners can enjoy financial savings, environmental advantages, and the satisfaction of contributing to a greener world, inspiring others to follow suit and promote water conservation and environmental stewardship.

7

MAINTAINING AND MONITORING YOUR RAINWATER HARVESTING SYSTEM

A farmer collects water from a well in the middle of his lush green field.

Congratulations on taking the first step towards a more sustainable and eco-friendly lifestyle by installing a rainwater harvesting system for your home! As you reap the benefits of this environmentally responsible decision, you must understand the importance of maintaining and monitoring your system to ensure its longevity and optimal performance. This chapter will provide the necessary knowledge and tools to keep your rainwater harvesting system in excellent condition, allowing you to enjoy its benefits for years.

A well-designed and properly installed rainwater harvesting system can provide a **reliable and sustainable water source** for various household uses, such as irrigation, toilet flushing, and even drinking water, if properly treated. However, like any other system, it requires regular maintenance and monitoring to function efficiently and provide clean, safe water for your home. This chapter will guide you through the essential steps to keep your system in top shape, from regular inspection and cleaning of gutters and downspouts to monitoring water quality and system efficiency.

By following the recommendations and best practices outlined in this chapter, you will ensure the smooth operation of your rainwater harvesting system, protect your investment, and contribute to a healthier environment. So, let's dive into maintaining and monitoring your rainwater harvesting system and learn how to make the most of this valuable resource.

Regular Inspection and Cleaning of Gutters and Downspouts

A well-maintained rainwater harvesting system is crucial for ensuring optimal performance and longevity. One of the critical components in this process is the regular inspection and cleaning of gutters and downspouts. This section will discuss the importance of keeping these components clean, how often they should be inspected, and the best practices for maintaining them.

Gutters and downspouts play a vital role in the rainwater harvesting process. They collect and channel rainwater from your roof to the storage tanks, ensuring it is efficiently captured and stored for future use. However, over time, debris such as leaves, twigs, and dirt can accumulate in your gutters and downspouts, causing blockages and reducing the efficiency of your rainwater harvesting system. Therefore, **regular inspection and cleaning of these components are essential** to prevent such issues and maintain the overall performance of your system.

It is recommended that you inspect your gutters and downspouts at least twice a year, ideally during the spring and fall seasons when debris accumulation is most likely to occur. However, if you live in an area with heavy foliage or frequent storms, you may need to inspect and clean your gutters more frequently. Look for any signs of blockages, such as overflowing gutters during rainfall or water pooling around the base of your downspouts.

When it comes to cleaning your gutters and downspouts, safety should be your top priority. Use a sturdy ladder to access your gutters and ensure it is properly secured before you begin. It is also a good idea to wear gloves and safety goggles to protect yourself from any debris or insects that may be present.

Begin by removing large debris, such as leaves and twigs, from your gutters or with a gutter scoop. Next, use a garden hose with a high-pressure nozzle to remove any remaining debris and dirt from the gutters and downspouts. Finally, direct the water flow away from your home's foundation to prevent potential water damage.

While cleaning your gutters and downspouts, please take the opportunity to inspect them for any signs of damage, such as cracks, rust, or loose connections. If you notice any issues, it is essential to address them promptly to prevent further damage to your rainwater harvesting system.

In conclusion, regular inspection and cleaning of your gutters and downspouts are essential for maintaining the efficiency and longevity of your rainwater harvesting system. By keeping these components clean and in good working order, you can ensure that your system continues to provide you with a reliable source of water for years to come.

Checking and Maintaining Storage Tanks and Filters

Proper maintenance of storage tanks and filters is crucial to your rainwater harvesting system's overall performance and longevity. This section will discuss the importance of regular inspections, cleaning, and maintenance of these components to ensure that your system continues to provide clean and safe water for your household needs.

Storage tanks are the heart of your rainwater harvesting system, storing the collected water until needed. Unfortunately, over time, sediment, debris, and algae can accumulate in the tanks, affecting water quality and even damaging the tank itself. To prevent these issues, inspecting and maintaining your storage tanks regularly is essential.

Begin by visually inspecting the tank for any signs of damage, such as cracks, leaks, or corrosion. If you notice any issues, you must address them immediately to prevent further damage or contamination of your water supply. Additionally, check for any signs of algae growth, indicating inadequate sunlight protection or poor water circulation. If algae are present, you may need to clean the tank and consider adding a UV light or other algae control measures.

To clean your storage tank:

1. Drain the water and remove any sediment or debris settled at the bottom. You can use a wet-dry vacuum or a specialized tank cleaning tool to accomplish this task.
2. Scrub the tank's interior surfaces with a long-handled brush and a mild, non-toxic cleaning solution.
3. Rinse the tank thoroughly with clean water and allow it to dry before refilling.

Filters play a vital role in maintaining the quality of your harvested rainwater by removing debris, sediment, and other contaminants before the water enters your storage tank. Regular inspection and cleaning of your filters are necessary to ensure their effectiveness and prevent clogging.

First, inspect the filters for any signs of damage or wear, such as tears or holes in the mesh or filter media. If you find any issues, replace the damaged components immediately. Next, remove the filter media and clean it according to the manufacturer's instructions. This may involve rinsing, brushing, or soaking the media in a cleaning solution. Finally, replace the filter media if it shows signs of wear or if the manufacturer recommends replacement after a certain period.

Finally, check the filter housing and connections for any signs of leaks or damage. Tighten any loose connections and replace any damaged components as needed. Regularly inspecting and maintaining your storage tanks and filters ensure that your rainwater harvesting system continues to provide clean, safe water for your home while minimizing the risk of system failure or damage.

Monitoring Water Quality and System Efficiency

Ensuring the quality of the harvested rainwater and the overall efficiency of your rainwater harvesting system is crucial for the success of your sustainable water management efforts. In this section, we will discuss the importance of monitoring water quality and system efficiency and provide practical tips and guidelines to help you keep your rainwater harvesting system in optimal condition.

One of the primary reasons for monitoring water quality is to ensure that the harvested rainwater is **safe for its intended use.** Whether you plan to use the water for irrigation, household cleaning, or even for drinking, it is essential to regularly test the water quality to avoid potential health risks and maintain the effectiveness of your system.

To monitor water quality, you can start by visually inspecting the water for any signs of contamination, such as discoloration, cloudiness, or debris. Additionally, you can use water testing kits available in the market to check for specific contaminants like bacteria, heavy metals, and chemicals. These kits usually come with easy-to-follow instructions and provide quick results, allowing you to take immediate action if necessary.

Another essential aspect of monitoring your rainwater harvesting system is evaluating its efficiency. This involves assessing the performance of various components, such as gutters, downspouts, filters, and storage tanks, to ensure they function optimally. In addition, regularly inspecting these components for signs of wear, damage, or blockage can help you identify potential issues early on and address them before they escalate.

To gauge the efficiency of your system, consider tracking the amount of water you collect and use over time. This can help you identify any discrepancies between the expected and actual water yield, which may indicate issues with your system. Additionally, monitoring your water usage can help you make informed decisions about future system upgrades or adjustments to improve efficiency.

In summary, monitoring water quality and system efficiency is essential to maintaining your rainwater harvesting system. By regularly inspecting and testing the water and evaluating the performance of various system components, you can ensure the longevity and optimal performance of your rainwater harvesting system. This contributes to a more sustainable water management approach and helps you enjoy the numerous benefits of rainwater harvesting for years to come.

Troubleshooting Common Issues and Seeking Professional Help

As with any home system, your rainwater harvesting setup may occasionally encounter issues requiring troubleshooting or professional assistance. In this section, we will discuss some common problems you may face and provide guidance on when to call in the experts.

One common issue that homeowners may experience is a **decrease in water pressure.** Various factors, such as clogged filters, blocked pipes, or malfunctioning pumps, can cause this. If you notice a drop in water pressure, check your filters and clean or replace them as necessary. If the issue persists, inspect your pipes for blockages and ensure your pump functions correctly. If you cannot identify or resolve the problem, it may be time to consult a professional.

Another issue that may arise is a **leak in your storage tank or pipes.** Regularly inspect your tank and pipes for any signs of leakage, such as damp spots, mold, or pooling water. If you discover a leak, it is crucial to address it promptly to prevent water loss and potential damage to your property. Small leaks may be repairable with a waterproof sealant, but more significant leaks or cracks may necessitate professional intervention.

Water quality is a crucial aspect of your rainwater harvesting system, and any changes in taste, odor, or appearance should be taken seriously. If you notice any unusual changes in your harvested water, it is essential to test it for contaminants. You can purchase water testing kits online or from local home improvement stores. If the test results indicate the presence of harmful substances, consult a professional to determine the appropriate course of action.

Sometimes, your rainwater harvesting system **may collect less water than you anticipated.** This could be due to issues with your catchment area, such as insufficient roof surface area or poor gutter design. It could also result from leaks or evaporation from your storage tank. Suppose you suspect that your system is underperforming. In that case, consulting with a professional who can assess your setup and recommend improvements may be worthwhile.

Finally, regular maintenance is critical to preventing many of these common issues. By staying vigilant and addressing any problems as they arise, you can help ensure your rainwater harvesting system's longevity and optimal performance. However, if you ever feel unsure about handling a particular issue or if your system requires more extensive repairs, feel free to seek the help of a professional. Their expertise can save you time, money, and frustration in the long run and help you maintain a safe and efficient rainwater harvesting system for your home.

Ensuring Longevity and Optimal Performance of Your Rainwater Harvesting System

In conclusion, diligent maintenance and monitoring is the key to ensuring the longevity and optimal performance of your rainwater harvesting system. By taking the time to regularly inspect, clean, and maintain the various components of your system, you can prevent potential issues from escalating and ensure that your system continues to function efficiently.

The gutters and downspouts are crucial in directing rainwater to your storage tanks. Keeping them clean and free from debris will improve the quality of the harvested water and prevent damage to your roof and home structure. Similarly, regularly checking and maintaining your storage tanks and filters will help to preserve water quality and extend the life of your system.

Monitoring water quality and system efficiency is essential to guarantee that your rainwater harvesting system provides a reliable and safe water source. By conducting periodic tests and keeping an eye on the overall performance of your system, you can identify any issues early on and take the necessary steps to address them.

If you encounter any problems or concerns with your rainwater harvesting system, feel free to seek professional help. A qualified expert can provide valuable guidance and assistance in troubleshooting and resolving any issues that may arise.

By following the tips and advice outlined in this chapter, you can take pride in knowing that you are contributing to a more sustainable and environmentally friendly way of life and enjoying the numerous benefits of having a well-maintained and efficient rainwater harvesting system. With proper care and attention, your system will continue to serve you and your home for many years.

Chapter Summary

1. Regular maintenance and monitoring of your rainwater harvesting system are crucial for ensuring its longevity and optimal performance.
2. Inspect and clean gutters and downspouts at least twice a year to prevent blockages and maintain the efficiency of your system.
3. Regularly check storage tanks and filters for signs of damage, wear, or blockage, and clean or replace them as necessary to maintain water quality and system performance.
4. Monitor water quality by visually inspecting the water and using water testing kits to check for contaminants, ensuring the harvested rainwater is safe for its intended use.
5. Evaluate system efficiency by assessing the performance of various components and tracking the amount of water collected and used over time.
6. Troubleshoot common issues, such as decreased water pressure or leaks, and seek professional help when necessary to address more complex problems.
7. Regular maintenance can prevent many common issues and help ensure your rainwater harvesting system's longevity and optimal performance.
8. By following the tips and advice outlined in this chapter, you can contribute to a more sustainable and environmentally friendly way of life while enjoying the numerous benefits of a well-maintained and efficient rainwater harvesting system.

8

UTILIZING HARVESTED RAINWATER FOR VARIOUS HOUSEHOLD PURPOSES

A rain barrel overflowing with water, surrounded by lush green plants.

Rainwater harvesting is an ancient practice utilized by various civilizations throughout history. However, it is gaining **renewed attention** as a sustainable and environmentally friendly solution to address water scarcity and reduce our reliance on municipal water supplies. This chapter will explore the numerous ways harvested rainwater can be used within your home, making it an invaluable resource for daily living.

Rainwater is a versatile and abundant natural resource that can be collected, stored, and utilized for various household purposes. By implementing a rainwater harvesting system in your home, you can significantly reduce your water consumption, lower your utility bills, and contribute to conserving precious water resources. Moreover, using rainwater for various household tasks can help you maintain a healthier and more eco-friendly lifestyle.

The following sections will explore how to use harvested rainwater within your home, from irrigation and gardening to indoor cleaning and laundry applications. We will also discuss how to safely use rainwater for toilet flushing, bathroom use, cooking, and drinking, ensuring that you are well-equipped to maximize the benefits of your rainwater harvesting system.

As you read through this chapter, you will gain valuable insights into the practical applications of harvested rainwater and learn how to integrate it effectively into your daily routines. By the end, you will be inspired to harness the power of rainwater and transform your home into a more sustainable and eco-conscious haven. So, let's embark on this exciting journey and discover how harvested rainwater can enhance your household and improve your quality of life.

Irrigation and Gardening with Rainwater

Rainwater harvesting is an environmentally friendly practice and a practical solution for homeowners looking to reduce their water bills and maintain a lush, thriving garden. In this section, we will explore the numerous benefits of using harvested rainwater for irrigation and gardening purposes and provide tips on effectively utilizing this valuable resource.

One of the primary advantages of using harvested rainwater for irrigation is that it is **free of chemicals and minerals** commonly found in tap water. These substances can harm plants, as they may cause a buildup of salts in the soil, hindering plant growth and overall health. Rainwater, on the other hand, is naturally soft and free of these contaminants, making it an ideal choice for nourishing your garden.

In addition to being chemical-free, rainwater is an excellent source of nutrients for plants. As rainwater passes through the atmosphere, it collects nitrogen, a vital nutrient for plant growth. So when you use harvested rainwater for irrigation, you provide your plants with a natural fertilizer, promoting more robust and healthier growth.

To effectively utilize harvested rainwater for irrigation, it is essential to have a proper storage system in place. Rain barrels or cisterns can collect and store rainwater, which can then be distributed throughout your garden using a hose, drip irrigation system, or a simple watering can. When setting up your rainwater harvesting system, please place it in a location that allows easy access to your garden. Consider the size of your storage container based on your garden's water needs.

When using rainwater for irrigation, monitoring your garden's moisture levels is crucial to avoid overwatering. While rainwater is a valuable resource, providing your plants with too much water is still possible, which can lead to root rot and other issues. To maintain optimal moisture levels, consider investing in a soil moisture meter or use your finger to gauge the soil's dampness before watering.

In conclusion, utilizing harvested rainwater for irrigation and gardening is a sustainable and cost-effective way to maintain a healthy and vibrant garden. By providing your plants with chemical-free, nutrient-rich water, you promote their growth and contribute to a greener and more sustainable future. With the right storage system and proper watering techniques, you can maximize the benefits of rainwater harvesting and enjoy a flourishing garden all year round.

Indoor Cleaning and Laundry Applications

As we continue to explore the various ways harvested rainwater can be utilized within your home, let's delve into the world of indoor cleaning and laundry applications. By using rainwater for these tasks, you save on your water bill and contribute to a more sustainable and eco-friendly lifestyle. This section will discuss how rainwater can be effectively used for cleaning and laundry purposes, as well as the necessary precautions to ensure optimal results.

One of the most significant advantages of using rainwater for cleaning and laundry is its **softness.** Unlike hard water, which contains high levels of minerals like calcium and magnesium, rainwater is naturally soft. This means it is less likely to leave residue on surfaces, making it an excellent choice for cleaning tasks such as mopping floors, washing windows, and wiping down countertops. The absence of minerals in rainwater also means that it is gentle on your appliances, helping to prolong their lifespan and maintain their efficiency.

When it comes to laundry, using rainwater can have several benefits. Firstly, the softness of the water means that you can use less detergent, as it will lather more easily. This saves you money and reduces the chemicals released into the environment. Additionally, the lack of minerals in rainwater prevents the buildup of limescale in your washing machine, which can lead to reduced efficiency and costly repairs. Finally, clothes washed in rainwater tend to feel softer. They look brighter, as no minerals can dull the colors or leave a residue on the fabric.

To use harvested rainwater for indoor cleaning and laundry applications, it is essential to **ensure that the water is adequately filtered and debris-free.** While rainwater is generally clean, it can still pick up contaminants from the atmosphere or your roof and gutters. Therefore, it is crucial to have a proper filtration system in place to remove any particles and impurities. Additionally, using a separate pump and plumbing system for your rainwater supply is recommended to avoid cross-contamination with your main water supply.

In conclusion, utilizing harvested rainwater for indoor cleaning and laundry applications is an excellent way to save money, reduce your environmental impact, and enjoy the benefits of naturally soft water. By taking the necessary precautions and ensuring that your rainwater is adequately filtered, you can enjoy a cleaner, greener, and more efficient home.

Rainwater for Toilet Flushing and Bathroom Use

One of the most practical and efficient ways to utilize harvested rainwater in your home is for toilet flushing and bathroom use. Doing so can significantly reduce your household's potable water consumption and contribute to a more sustainable and eco-friendly lifestyle. In this section, we will explore the benefits of using rainwater for toilet flushing and bathroom use and provide guidance on how to set up a system for this purpose.

Toilet flushing accounts for a **substantial portion of a household's water usage**, with an average of 30% of indoor water consumption attributed to this activity. Therefore, harvesting rainwater can save potable water and reduce your water bill. Additionally, using rainwater for toilet flushing can help alleviate pressure on local water resources and reduce the energy required for water treatment and distribution.

Setting up a rainwater harvesting system for toilet flushing and bathroom use is relatively simple. First, you will need a storage tank to collect the rainwater. This tank should be connected to your home's gutter system, directing rainwater into the tank. Then, install a filter at the entrance of the tank to prevent debris and contaminants from entering the stored water.

Next, you will need a pump and a separate plumbing system to distribute the harvested rainwater to your toilets and bathroom fixtures. This system should be independent of your potable water supply to avoid cross-contamination. Ensuring the pump is powerful enough to provide adequate water pressure for your bathroom needs is also crucial.

To further enhance the water quality of the harvested rainwater, you can install a secondary filter or treatment system, such as a UV sterilizer or a chlorine dosing system. This step is crucial if you use rainwater for showering or bathing. In addition, it will help eliminate any potential health risks associated with using untreated rainwater.

In conclusion, using harvested rainwater for toilet flushing and bathroom use is an excellent way to conserve potable water and reduce your environmental footprint. By implementing a rainwater harvesting system in your home, you can enjoy the benefits of a more sustainable lifestyle while saving money on your water bill. In addition, with proper planning and installation, you can ensure that your rainwater harvesting system is safe, efficient, and effective in meeting your household's bathroom needs.

Cooking and Drinking: Ensuring Water Quality and Safety

As a homeowner who has embraced rainwater harvesting, you have already taken a significant step towards sustainable living and reducing your water footprint. While harvested rainwater can be used for various household purposes, it is essential to ensure its quality and safety, especially regarding cooking and drinking. In this section, we will discuss the necessary steps to guarantee that your harvested rainwater meets the required standards for consumption and culinary use.

First and foremost, it is crucial to maintain the cleanliness of your rainwater harvesting system. Regular inspection and cleaning of your catchment area, gutters, filters, and storage tanks will prevent the accumulation of debris, algae, and other contaminants that could compromise water quality. Additionally, it is advisable to install first-flush diverters to discard the initial flow of rainwater, which may contain higher levels of pollutants from the catchment surface.

Once you have ensured the cleanliness of your system, the next step is to treat the harvested rainwater to make it safe for consumption. Depending on your specific needs and preferences, several methods are available for treating rainwater. Some popular options include:

Filtration: Installing a multi-stage filtration system can effectively remove sediment, debris, and other suspended particles from the harvested rainwater. This process typically involves using sediment filters, activated carbon filters, and ultrafiltration or reverse osmosis membranes.

Disinfection: To eliminate harmful microorganisms such as bacteria, viruses, and protozoa, disinfection methods like chlorination, ultraviolet (UV) light treatment, or ozonation. These processes are highly effective in ensuring that your harvested rainwater is free from pathogens and safe for consumption.

Boiling: As a simple and cost-effective method, boiling your harvested rainwater for at least one minute can kill harmful microorganisms. However, boiling does not remove chemical contaminants or improve the taste and odor of the water.

Water testing: Regularly testing your harvested rainwater for contaminants is essential to ensure its safety for cooking and drinking. You can use home water testing kits or send samples to a certified laboratory for a comprehensive analysis. Based on the test results, you can adjust your treatment methods accordingly.

Once you have treated your harvested rainwater, it is ready for use in cooking and drinking. Using rainwater for these purposes reduces your reliance on municipal water supplies. In addition, it gives you a sense of satisfaction to know you are consuming a sustainable and eco-friendly resource. By following the guidelines discussed in this section, you can maximize the benefits of rainwater harvesting in your home while ensuring the health and well-being of your family.

Maximizing the Benefits of Rainwater Harvesting in Your Home

In conclusion, rainwater harvesting is an environmentally-friendly and cost-effective solution to address water scarcity and reduce the strain on our planet's limited resources. Implementing a rainwater harvesting system in your home can significantly decrease your dependence on municipal water supplies and contribute to a more sustainable future.

Throughout this chapter, we have explored various household applications for harvested rainwater, including irrigation and gardening, indoor cleaning and laundry, toilet flushing, bathroom use, cooking, and drinking. Each application offers unique benefits and opportunities for homeowners to maximize their harvested rainwater.

To maximize the benefits of rainwater harvesting in your home, it is essential to consider the following key points:

Choose the right system: Select a rainwater harvesting system that best suits your needs and budget. This may include factors such as the size of your property, the amount of rainfall in your region, and your specific water usage requirements.

Maintain your system: Regular maintenance of your rainwater harvesting system is crucial to ensure its efficiency and longevity. This includes cleaning gutters and filters, inspecting storage tanks, and monitoring water quality.

Optimize water usage: Be mindful of your water consumption habits and use harvested rainwater efficiently. This may involve using water-saving devices like drip irrigation systems and low-flow toilets and adopting water conservation practices in your daily routines.

Ensure water quality and safety: If you plan to use harvested rainwater for cooking and drinking, it is vital to implement appropriate treatment and filtration methods to guarantee the water's safety and quality. In addition, regular water supply testing and monitoring can help you maintain the highest water quality standards.

Educate and inspire others: Share your experiences and knowledge about rainwater harvesting with your friends, family, and community. By raising awareness about the benefits of rainwater harvesting, you can inspire others to adopt this sustainable practice and contribute to a greener future.

In summary, rainwater harvesting offers many benefits for homeowners, from reducing water bills to promoting self-sufficiency and environmental stewardship. By utilizing harvested rainwater for various household purposes and adopting best practices, you can maximize the potential of your rainwater harvesting system and enjoy the rewards of this sustainable solution for years to come.

Chapter Summary

1. Rainwater harvesting is an ancient practice that has gained renewed attention as a sustainable and environmentally friendly solution to address water scarcity and reduce reliance on municipal water supplies.
2. Harvested rainwater can be used for various household purposes, including irrigation and gardening, indoor cleaning and laundry, toilet flushing and bathroom use, and cooking and drinking, provided it is treated and filtered correctly.
3. Rainwater is naturally soft and free of chemicals and minerals, making it an ideal choice for healthy plants and maintaining a healthy garden.
4. Using rainwater for indoor cleaning and laundry applications can save money, reduce environmental impact, and prolong the lifespan of appliances.
5. Implementing a rainwater harvesting system for toilet flushing and bathroom use can significantly reduce household water consumption and contribute to a more sustainable lifestyle.
6. Ensuring the quality and safety of harvested rainwater for cooking and drinking is crucial. It can be achieved through proper system maintenance, filtration, disinfection, and regular water testing.
7. Maximizing the benefits of rainwater harvesting in your home involves choosing the right system, maintaining it, optimizing water usage, ensuring water quality and safety, and educating and inspiring others.
8. By adopting rainwater harvesting practices and utilizing harvested rainwater for various household purposes, homeowners can contribute to a greener future and enjoy the rewards of this sustainable solution.

9

LEGAL AND REGULATORY CONSIDERATIONS FOR RAINWATER HARVESTING

A rain barrel overflowing with water, surrounded by lush green plants.

Rainwater harvesting is environmentally friendly and cost-effective to conserve water, reduce your reliance on municipal water supplies, and lower utility bills. However, as more and more homeowners recognize the benefits of rainwater harvesting, they must be aware of the **legal and regulatory considerations** of implementing such systems at home. This chapter aims to provide you with a comprehensive understanding of the various rules and regulations governing rainwater harvesting, ensuring that your system is compliant and sustainable.

While rainwater harvesting has been practiced for centuries, modern-day systems are subject to various laws and regulations depending on your location. These rules are in place to protect public health, ensure your home's safety and structural integrity, and prevent potential conflicts with neighbors or homeowners associations. By familiarizing yourself with these legal and regulatory considerations, you can avoid fines, disputes, and other issues arising from non-compliance.

This chapter will explore the different aspects of legal and regulatory compliance for rainwater harvesting systems, including understanding local and state regulations, obtaining necessary permits and licenses, complying with building codes and standards, and navigating homeowners association rules and restrictions. By the end of this chapter, you will have a solid foundation of knowledge to help you ensure that your rainwater harvesting system is both legally compliant and environmentally sustainable.

Understanding Local and State Regulations

Rainwater harvesting is a sustainable and eco-friendly practice that has recently gained significant popularity. However, before embarking on your rainwater harvesting journey, it is crucial to understand the legal landscape that governs this practice. Local and state regulations significantly determine the feasibility and legality of rainwater harvesting for your home. This section will explore the importance of familiarizing yourself with these regulations and guide how to navigate them effectively.

First and foremost, it is essential to recognize that rainwater harvesting regulations can vary greatly depending on your location. Some states and municipalities have embraced this practice and actively encouraged it through incentives and lenient regulations. On the other hand, certain areas may have more stringent rules or even prohibit rainwater harvesting altogether. This variation in regulations underscores the importance of researching and understanding the specific laws that apply to your area.

To begin your research, consult your local government's website or contact the relevant department, such as the Department of Environmental Protection or Water Resources. These sources can provide valuable information on the legalities of rainwater harvesting in your area and any specific requirements or restrictions that may apply. It may also be helpful to consult with local experts or organizations specializing in rainwater harvesting, as they can offer firsthand knowledge and experience in navigating the regulatory landscape.

When researching local and state regulations, pay particular attention to the following aspects:

Legal status of rainwater harvesting: Determine whether rainwater harvesting is allowed in your area and if there are any restrictions on its use, such as for potable or non-potable purposes.

Rainwater collection methods: Some jurisdictions may have specific requirements or restrictions on the types of collection systems used, such as roof catchment systems or ground-level collection.

Water rights and ownership: In certain areas, water rights may be a complex issue, and rainwater harvesting could infringe on the rights of other water users. Be sure to understand the implications of water rights in your area and how they may affect your rainwater harvesting plans.

Water quality and treatment: Specific water quality standards and treatment requirements may be met depending on the harvested rainwater's intended use. Please familiarize yourself with these standards and ensure your planned system meets them.

Stormwater management and runoff: Some jurisdictions may have regulations to manage stormwater runoff and prevent potential flooding or environmental issues. Ensure your rainwater harvesting system complies with these regulations and does not negatively impact stormwater management efforts.

By taking the time to thoroughly understand the local and state regulations that govern rainwater harvesting, you can ensure that your system is both legally compliant and environmentally sustainable. In addition, this knowledge will help you avoid potential fines or penalties and contribute to the broader goal of promoting responsible water use and conservation in your community.

Obtaining Necessary Permits and Licenses

As you embark on your journey to integrate rainwater harvesting into your home, you must understand the permits and licenses required to ensure a smooth and legally compliant process. This section will guide you through the steps to obtain the necessary permits and licenses for your rainwater harvesting system.

First and foremost, it is crucial to research and familiarize yourself with the local and state regulations governing rainwater harvesting in your area. These regulations may vary significantly from location to location, so gathering **accurate and up-to-date information** is essential. You can consult your local government's website, contact your city or county's permitting office, or visit your local library to access resources.

Once you clearly understand the regulations, you can obtain the required permits and licenses. This may involve submitting an application to your local permitting office, providing detailed plans and specifications of your proposed rainwater harvesting system, and paying any associated fees. Be prepared to answer questions about the system's design, capacity, intended use, and any safety measures you plan to implement.

Sometimes, you may also need approval from other agencies or organizations, such as your local health department or water utility company. These organizations may have additional requirements or guidelines for rainwater harvesting systems, particularly if you plan to use the harvested water for potable purposes. Therefore, it is essential to communicate with these entities early in the process to avoid any delays or complications.

As you work through the permitting process, be prepared to adjust your rainwater harvesting system's design or specifications to meet regulatory requirements. This may involve modifying the system's capacity, incorporating additional filtration or treatment components, or implementing specific safety measures. It is essential to remain flexible and open to feedback from regulatory agencies to ensure your system is compliant and effective.

Finally, once you have obtained all necessary permits and licenses, it is crucial to keep them readily available and accessible. This will allow you to demonstrate compliance with local and state regulations should any questions or concerns arise. Also, maintaining accurate records of your permits and licenses can be helpful when selling your home or applying for insurance coverage, demonstrating your commitment to sustainable and responsible water management practices.

In conclusion, obtaining the necessary permits and licenses for your rainwater harvesting system is critical in ensuring compliance with local and state regulations. By thoroughly researching the requirements, engaging with regulatory agencies, and maintaining accurate records, you can confidently move forward with your rainwater harvesting project, knowing that you are contributing to a sustainable and environmentally friendly solution for your home.

Complying with Building Codes and Standards

This section will explore the importance of adhering to building codes and standards when implementing a rainwater harvesting system for your home. By understanding and complying with these regulations, you can ensure your rainwater harvesting system's safety, efficiency, and longevity.

Building codes and standards are established to provide a uniform set of guidelines for designing, constructing, and maintaining buildings and their various systems. These regulations are in place to ensure the safety and well-being of the occupants and protect the environment and surrounding community. Regarding rainwater harvesting, building codes and standards typically address the materials used, system design, installation, and maintenance.

To begin, it is essential to familiarize yourself with the specific building codes and standards for rainwater harvesting in your area. These regulations may vary depending on your location, so it is crucial to consult with your local building department or a knowledgeable professional to obtain accurate information. Additionally, it is essential to note that building codes and standards are subject to change, so staying up-to-date on any revisions is critical to maintaining compliance.

When designing your rainwater harvesting system, be sure to consider the following aspects that building codes and standards may regulate:

Materials: The materials used in your rainwater harvesting system must be durable, non-toxic, and suitable for their intended purpose. This may include using specific types of pipes, fittings, tanks, and filters that meet established standards.

System Design: The overall design of your rainwater harvesting system should adhere to guidelines that ensure its functionality and safety. This may involve requirements related to storage tanks' size and location, plumbing connections' configuration, and the incorporation of overflow and backflow prevention measures.

Installation: Proper installation is crucial to the performance and longevity of your rainwater harvesting system. Building codes and standards may dictate specific installation methods and require inspections and approval from local authorities.

Maintenance: Regular maintenance is necessary to keep your rainwater harvesting system operating efficiently and safely. Building codes and standards may outline recommended maintenance procedures and require periodic inspections and reporting to local authorities.

By complying with building codes and standards, you can ensure that your rainwater harvesting system is safe and efficient and contributes to the sustainable management of water resources in your community. The following section will discuss navigating homeowners association rules and restrictions that may impact your rainwater harvesting project.

Navigating Homeowners Association Rules and Restrictions

As you embark on your rainwater harvesting journey, you must be aware of and adhere to the rules and restrictions set forth by your local **Homeowners Association (HOA).** These organizations often have specific guidelines that govern the appearance and functionality of homes within their communities. In this section, we will discuss the importance of understanding your HOA's rules and restrictions and how to navigate them to ensure a successful and compliant rainwater harvesting system installation.

First and foremost, it is crucial to familiarize yourself with your HOA's governing documents, which typically include the Covenants, Conditions, and Restrictions (CC&Rs) and any additional rules and regulations specific to your community. These documents outline the standards and requirements for various aspects of your property, such as landscaping, exterior modifications, and sustainability initiatives like rainwater harvesting.

Consider any provisions impacting your rainwater harvesting plans when reviewing your HOA's rules and restrictions. For example, there may be guidelines regarding the size, location, and appearance of rain barrels or cisterns and requirements for screening or concealing these components from the public view. Additionally, your HOA may have specific rules about the materials that can be used for your rainwater harvesting system, such as requiring non-toxic, food-grade materials for water storage containers.

Suppose your HOA's rules and restrictions must be clarified or specifically address rainwater harvesting. In that case, contacting your HOA board or management company for clarification is a good idea. They can guide any existing policies or help you understand how your proposed system might fit within the community's overall aesthetic and sustainability goals. Sometimes, you may need to submit a formal request or application to your HOA to approve your rainwater harvesting system. This process may require you to provide detailed information about your proposed system, including its design, materials, and intended use.

When working with your HOA, it is essential to maintain open lines of communication and demonstrate a willingness to collaborate. By being proactive and engaging in a constructive dialogue, you can educate your HOA about the benefits of rainwater harvesting and potentially pave the way for future installations within your community. Be prepared to address any concerns or questions. Consider sharing resources or examples of successful rainwater harvesting systems in other HOA-governed communities.

In conclusion, navigating your HOA's rules and restrictions is critical in implementing a rainwater harvesting system for your home. By understanding and adhering to these guidelines, you can ensure that your system is compliant and contributes to the overall sustainability and aesthetic appeal of your community. In addition, with careful planning and open communication, you can successfully integrate rainwater harvesting into your home while maintaining a positive relationship with your HOA.

Ensuring Compliance and Sustainable Rainwater Harvesting

In conclusion, the journey towards implementing a rainwater harvesting system for your home is not only about the technical aspects and environmental benefits but also about understanding and adhering to legal and regulatory considerations. By taking the time to familiarize yourself with local and state regulations, obtaining the necessary permits and licenses, complying with building codes and standards, and navigating homeowners association rules and restrictions, you can ensure that your rainwater harvesting system is both compliant and sustainable.

As a **responsible homeowner,** it is essential to recognize the importance of these legal and regulatory aspects, as they are in place to protect the environment, public health, and safety. Moreover, they help promote the responsible use of water resources and encourage adopting sustainable practices.

By following the guidelines and recommendations outlined in this chapter, you can confidently embark on your rainwater harvesting journey, knowing that you are taking the necessary steps to ensure compliance with the law and contribute to a more sustainable future for your community and the planet.

Remember, setting up a rainwater harvesting system for your home is not a one-time event but an ongoing commitment to sustainability and resource conservation. As you maintain and optimize your system, stay informed about any changes in regulations and best practices to ensure your efforts remain compliant and effective.

In the end, the rewards of rainwater harvesting go beyond the satisfaction of contributing to a greener environment and reducing your water bills. By taking the time to understand and adhere to legal and regulatory considerations, you also foster a sense of community and shared responsibility, inspiring others to follow in your footsteps and embrace sustainable living practices.

So, as you embark on this exciting and rewarding journey, remember that your efforts to harvest rainwater are beneficial for your home and wallet and play a crucial role in promoting a more sustainable and water-conscious future for generations to come.

Chapter Summary

1. Rainwater harvesting is an environmentally friendly and cost-effective way to conserve water. Still, knowing the legal and regulatory considerations of implementing such systems at home is essential.
2. Local and state regulations significantly determine the feasibility and legality of rainwater harvesting for your

home. These regulations can vary greatly depending on your location.

3. Obtaining the necessary permits and licenses for your rainwater harvesting system is critical in ensuring compliance with local and state regulations.
4. Building codes and standards are established to provide a uniform set of guidelines for designing, constructing, and maintaining buildings and their various systems, including rainwater harvesting systems.
5. Homeowners Associations (HOAs) often have specific rules and restrictions that govern the appearance and functionality of homes within their communities, which may impact your rainwater harvesting plans.
6. Navigating your HOA's rules and restrictions is critical in implementing a rainwater harvesting system for your home. Maintaining open lines of communication with your HOA can help ensure a successful and compliant installation.
7. Ensuring compliance with legal and regulatory considerations is about protecting the environment, public health, and safety, promoting responsible water use, and adopting sustainable practices.
8. Setting up a rainwater harvesting system for your home is an ongoing commitment to sustainability and resource conservation. Staying informed about regulations and best practice changes is essential to ensure your efforts remain compliant and effective.

10

THE ENVIRONMENTAL AND ECONOMIC BENEFITS OF RAINWATER HARVESTING

A rain barrel overflowing with water, surrounded by lush green plants.

In today's rapidly changing world, where environmental concerns and resource management are on our minds, rainwater harvesting has emerged as a practical and sustainable solution for homeowners. This **age-old practice** of collecting and storing rainwater for later use has been gaining popularity in recent years as more people recognize its numerous environmental and economic benefits. By embracing rainwater harvesting, homeowners can contribute to a greener future and enjoy significant cost savings on their water bills.

Rainwater harvesting systems can range from simple rain barrels to more complex setups involving pumps, filters, and storage tanks. Regardless of the system's complexity, the core principle remains: capturing rainwater that would otherwise be lost as runoff and putting it to good use. This chapter will explore the various environmental and economic benefits of rainwater harvesting, including reducing the strain on municipal water supplies, minimizing the impact of stormwater runoff and soil erosion, promoting sustainable landscaping and gardening practices, saving money on water bills, and reducing energy consumption.

As you read through this chapter, you will gain a deeper understanding of how rainwater harvesting can positively impact both the environment and your wallet. Investing in a rainwater harvesting system for your home makes a wise financial decision. In addition, it takes a significant step towards a more sustainable and eco-friendly lifestyle. So, let's dive in and explore the myriad advantages of rainwater harvesting for your home and the environment.

Reducing the Strain on Municipal Water Supplies

As the world's population grows, the demand for clean, accessible water has increased. As a result, municipal water supplies are often stretched to their limits, struggling to meet the needs of both residential and commercial consumers. Homeowners can play a crucial role in alleviating this strain on our precious water resources by embracing rainwater harvesting.

Rainwater harvesting systems collect and store rainwater that would otherwise flow into storm drains, eventually returning to rivers, lakes, and oceans. By capturing this water, **homeowners can reduce their reliance on municipal water supplies**, freeing up resources for other essential services and uses. This is particularly important in areas that experience frequent droughts or water shortages, where every drop of water saved can make a significant difference.

The benefits of reducing the strain on municipal water supplies extend beyond the immediate water availability. As more people adopt rainwater harvesting practices, the need for costly infrastructure projects, such as dams and reservoirs, may be reduced. This can save taxpayers money and minimize the environmental impact associated with large-scale water management projects.

Furthermore, decreasing the demand for municipal water supplies can help protect our natural water sources from over-extraction. Many rivers, lakes, and aquifers are being depleted at an alarming rate, leading to devastating consequences for ecosystems and communities that rely on these water sources. By using harvested rainwater for non-potable purposes, such as irrigation and toilet flushing, homeowners can contribute to preserving these vital resources for future generations.

In summary, rainwater harvesting offers a practical and effective way for homeowners to reduce their dependence on municipal water supplies. By doing so, they can help conserve water resources, protect the environment, and support the long-term sustainability of our water supply systems.

Minimizing the Impact of Stormwater Runoff and Soil Erosion

One of the most significant yet often overlooked benefits of rainwater harvesting is its ability to **minimize the impact of stormwater runoff and soil erosion.** As urbanization expands, natural landscapes are replaced with impervious surfaces like concrete and asphalt. As a result, rain cannot be absorbed when it falls on these surfaces, leading to increased stormwater runoff. This excess water can cause many environmental issues, including flooding, soil erosion, and pollution of nearby water sources. We can play a vital role in mitigating these problems by implementing rainwater harvesting systems in our homes.

Stormwater runoff is a significant concern in urban areas, as it can overload storm drains and sewer systems. This can result in localized flooding, causing damage to property and infrastructure. By capturing rainwater before it becomes runoff, we can significantly reduce the volume of water entering these systems. This helps prevent flooding and decreases the number of pollutants washed into our rivers, lakes, and oceans.

In addition to reducing the strain on stormwater management systems, rainwater harvesting can also help to minimize soil erosion. When rain falls on natural landscapes, it is absorbed by the soil and vegetation, which helps maintain the land's stability. However, when rain falls on impervious surfaces, it can wash away the topsoil, leading to erosion. This can have devastating consequences for both the environment and the economy, as it can lead to the loss of fertile land, damage to infrastructure, and increased sedimentation in waterways.

By capturing rainwater and using it to irrigate our gardens and landscapes, we can help maintain the soil's health and stability. This not only prevents erosion but also promotes the growth of vegetation, which can further help to absorb and filter rainwater. Moreover, using harvested rainwater for irrigation can reduce the demand for potable water, often sourced from rivers and lakes. This can help to maintain healthy ecosystems and protect the habitats of various plant and animal species.

In conclusion, rainwater harvesting offers a practical and effective solution for minimizing the impact of stormwater runoff and soil erosion. By capturing and utilizing rainwater, we can reduce the strain on our stormwater management systems, protect our water sources from pollution, and maintain the health and stability of our landscapes. As we continue to embrace the advantages of rainwater harvesting, we can look forward to a more sustainable and resilient future for our homes and the environment.

Promoting Sustainable Landscaping and Gardening Practices

As we continue to explore the environmental and economic benefits of rainwater harvesting, it's essential to highlight its role in promoting sustainable landscaping and gardening practices. With the increasing awareness of the importance of eco-friendly approaches, rainwater harvesting is vital for homeowners looking to create a greener environment around their homes.

Sustainable landscaping refers to designing, constructing, and maintaining outdoor spaces that minimize resource consumption, reduce waste, and preserve the natural environment. Gardening practices, on the other hand, focus on nurturing plants and vegetation in an eco-friendly manner. In addition, homeowners can significantly contribute to a healthier and more sustainable ecosystem by incorporating rainwater harvesting into these practices.

One of the primary ways rainwater harvesting supports sustainable landscaping is by providing a **natural and chemical-free water source for plants.** Unlike treated municipal water, rainwater is free from chemicals such as chlorine and fluoride, which can harm plants and soil health. This means that by using harvested rainwater, you are conserving water resources and promoting a healthier environment for your plants to thrive.

Moreover, rainwater harvesting enables homeowners to maintain their gardens and landscapes during water restrictions or droughts. This is particularly important in regions where water scarcity is a pressing issue. By having a rainwater harvesting system, you can continue caring for your plants without relying on limited municipal water supplies.

Another advantage of using harvested rainwater for landscaping and gardening is its positive impact on soil quality. Rainwater is naturally soft, meaning it has a lower mineral content than hard water. This characteristic of rainwater helps prevent the buildup of salts in the soil, which can be detrimental to plant growth and soil structure. As a result, using rainwater for irrigation promotes a healthier soil environment that encourages plant growth and reduces the need for chemical fertilizers.

Lastly, sustainable landscaping and gardening practices often involve the use of native plants that are well-adapted to local climate conditions. These plants typically require less water and maintenance than non-native species, making them an ideal choice for eco-conscious homeowners. By combining native plants with rainwater harvesting, you can create a low-maintenance and water-efficient landscape that is both beautiful and environmentally friendly.

In conclusion, rainwater harvesting plays a significant role in promoting sustainable landscaping and gardening practices. By providing a natural, chemical-free water source, reducing the strain on municipal water supplies, and supporting soil health, rainwater harvesting allows homeowners to create a greener and more sustainable environment around their homes. Embracing these practices benefits the environment and contributes to a healthier and more vibrant outdoor space for you and your family to enjoy.

Saving Money on Water Bills and Reducing Energy Consumption

One of the most tangible benefits of rainwater harvesting is the potential to **save money on your water bills.** As water becomes an increasingly scarce resource, its cost rises. Collecting and utilizing rainwater can significantly reduce your reliance on municipal water supplies and decrease your monthly expenses.

Rainwater is a free resource that falls from the sky, making it an attractive alternative to purchasing water from your local utility company. Installing a rainwater harvesting system allows you to capture this valuable resource for various household purposes, such as watering your garden, flushing toilets, and even washing clothes. In many cases, rainwater can be used as a direct substitute for tap water, allowing you to save money without sacrificing convenience or quality.

In addition to direct financial savings, rainwater harvesting can also help you reduce your energy consumption. Treating and distributing water requires a significant amount of energy, contributing to greenhouse gas emissions and climate change. Using rainwater instead of relying solely on municipal water supplies can help reduce the energy required to pump and treat water, thereby decreasing your carbon footprint.

Furthermore, using rainwater can help reduce the demand for energy-intensive water heating. Rainwater is typically cooler than tap water, requiring less energy to heat up for various household uses. This can result in lower energy bills and a reduced environmental impact.

In conclusion, rainwater harvesting offers a multitude of financial and environmental benefits. By capturing and utilizing this free and abundant resource, you can save money on your water bills, reduce energy consumption, and contribute to a more sustainable future.

Investing in a Greener Future with Rainwater Harvesting

In conclusion, investing in rainwater harvesting systems for your home is a **wise financial decision** and a significant step towards a greener and more sustainable future. As we have explored throughout this chapter, the environmental and economic benefits of rainwater harvesting are numerous and far-reaching.

By reducing the strain on municipal water supplies, homeowners who embrace rainwater harvesting contribute to conserving precious water resources and help alleviate the pressure on water treatment facilities. This, in turn, leads to a decrease in energy consumption and a reduction in greenhouse gas emissions.

Furthermore, rainwater harvesting systems are crucial in minimizing the impact of stormwater runoff and soil erosion. By capturing and storing rainwater, we can prevent the harmful effects of uncontrolled water flow on our ecosystems, including the destruction of habitats and the pollution of waterways.

In addition to these environmental benefits, rainwater harvesting promotes sustainable landscaping and gardening practices. Using harvested rainwater for irrigation allows homeowners to maintain lush and healthy gardens without relying on chemically treated municipal water. This benefits the plants and supports the overall health of the surrounding environment.

Lastly, the economic benefits of rainwater harvesting must be balanced. By reducing reliance on municipal water supplies, homeowners can save considerable money on their water bills. Moreover, the reduced energy consumption associated with water treatment and distribution translates to lower energy costs and a decreased carbon footprint.

In summary, rainwater harvesting is an investment in a greener future for both individual homeowners and the planet as a whole. By adopting this sustainable practice, we can collectively work towards a more environmentally responsible and economically viable future, ensuring future generations' health and well-being. So, take the first step towards a greener future and start reaping the rewards of rainwater harvesting today.

Chapter Summary

1. Rainwater harvesting is a practical and sustainable solution for homeowners, offering numerous environmental and economic benefits, such as reducing strain on municipal water supplies, minimizing stormwater runoff and soil erosion, promoting sustainable landscaping and gardening practices, and saving money on water bills and energy consumption.
2. By capturing and storing rainwater, homeowners can reduce their reliance on municipal water supplies, freeing up resources for other essential services and uses and potentially reducing the need for costly infrastructure projects like dams and reservoirs.
3. Rainwater harvesting helps minimize the impact of stormwater runoff and soil erosion by capturing rainwater before it becomes runoff, reducing the volume of water

entering stormwater management systems, and preventing the washing away of topsoil.

4. Sustainable landscaping and gardening practices can be supported by rainwater harvesting, as it provides a natural, chemical-free source of water for plants, maintains soil health, and allows for native plants that require less water and maintenance.
5. Rainwater harvesting can save homeowners money on their water bills by reducing their reliance on municipal water supplies and utilizing rainwater for various household purposes, such as watering gardens, flushing toilets, and washing clothes.
6. Energy consumption can be reduced through rainwater harvesting, as it decreases the energy required to pump and treat water. Using rainwater can also reduce the demand for energy-intensive water heating.
7. Rainwater harvesting contributes to a greener and more sustainable future by conserving water resources, reducing energy consumption, and promoting eco-friendly landscaping and gardening practices.
8. Investing in rainwater harvesting systems is a wise financial decision and a significant step towards a more environmentally responsible and economically viable future for both individual homeowners and the planet as a whole.

AFTERWORD

As we conclude this enlightening exploration into the world of rainwater harvesting, we must pause and reflect on our journey together. From understanding the basic principles of rainwater harvesting to delving into the intricacies of various systems, we have uncovered the immense potential of this simple yet powerful practice. By embracing rainwater harvesting for our homes, we are taking a step towards self-sufficiency and contributing to a more sustainable and environmentally conscious future.

Throughout this book, we have seen how **rainwater harvesting can profoundly impact our environment and communities**. By reducing the strain on our water resources and decreasing the demand for energy-intensive water treatment processes, we are actively participating in preserving our planet. Moreover, we have learned that rainwater harvesting can foster community and shared responsibility as neighbors come together to implement and maintain these systems.

In our exploration, we have also addressed the challenges and misconceptions surrounding rainwater harvesting. By dispelling myths and providing accurate information, we have empowered ourselves with the knowledge to make informed decisions and confidently advocate for the adoption of rainwater harvesting in our communities.

We have also discussed the crucial role that government, industry, and individuals must play in promoting rainwater harvesting. Through policy changes, incentives, and education, we can create an environment where rainwater harvesting becomes the norm rather than the exception. By working together, we can ensure that the benefits of rainwater harvesting are accessible to all, regardless of socioeconomic status or geographical location.

As we envision a world where rainwater harvesting is commonplace, we can begin to imagine the countless benefits that would follow. From healthier ecosystems to stronger communities, the widespread adoption of rainwater harvesting would undoubtedly contribute to a more sustainable and resilient future.

In the following sections, we will delve deeper into the impact of rainwater harvesting on our environment and communities, explore the challenges that must be overcome, and discuss the roles that various stakeholders must play in promoting this practice. Finally, we will share our thoughts on the path forward for rainwater harvesting and how it can contribute to a sustainable future for all.

The Impact of Rainwater Harvesting on Our Environment and Communities

As we reach the end of our journey exploring rainwater harvesting, it is essential to take a moment to reflect on the profound impact this practice can have on our environment and communities. Rainwater harvesting is not just a means to conserve water but a **powerful tool** that can contribute to a more sustainable and resilient future for all.

First and foremost, rainwater harvesting helps to alleviate the stress on our overburdened water resources. With increasing population growth, urbanization, and climate change, the demand for freshwater is rapidly outstripping the supply. By capturing and utilizing rainwater, we can reduce reliance on traditional water sources, such as rivers, lakes, and groundwater, which are already under immense pressure. This, in turn, helps to preserve these vital resources for future generations and protect the delicate ecosystems that depend on them.

Moreover, rainwater harvesting can play a significant role in mitigating the effects of climate change. As extreme weather events, such as floods and droughts, become more frequent and severe, the need for adaptive water management strategies becomes increasingly urgent. Rainwater harvesting systems can help buffer against these impacts by providing an alternative water source during times of scarcity and reducing the risk of flooding by slowing down and storing stormwater runoff.

In addition to its environmental benefits, rainwater harvesting can have a transformative effect on our communities. By promoting self-sufficiency and resilience, rainwater harvesting empowers individuals and neighborhoods to take control of their water supply and reduce their dependence on centralized water systems. This fosters a sense of pride and ownership and encourages a more responsible and mindful approach to water use.

Furthermore, rainwater harvesting can help to bridge the gap between water-rich and water-poor communities. In many parts of the world, access to clean, safe water remains a luxury rather than a fundamental human right. By implementing rainwater harvesting systems, these communities can tap into a previously underutilized resource and improve their water security, health, and overall quality of life.

Lastly, the widespread adoption of rainwater harvesting can stimulate local economies by creating new job opportunities in designing, installing, and maintaining these systems. This, in turn, can contribute to the growth of a green economy, which prioritizes environmental sustainability and social equity.

In conclusion, the impact of rainwater harvesting on our environment and communities is profound and far-reaching. Yet, by embracing this practice, we can work together to create a more sustainable, resilient, and equitable future for all.

Overcoming Challenges and Misconceptions Surrounding Rainwater Harvesting

As we have explored the world of rainwater harvesting throughout this book, addressing the challenges and misconceptions that often arise when discussing this sustainable practice is essential. By overcoming these hurdles, we can pave the way for a future where rainwater harvesting is widely adopted and appreciated for its numerous benefits.

One common misconception is that rainwater harvesting is a complicated and expensive process. While it is true that some systems can be quite ***intricate***, there are also many simple and affordable options available for homeowners. From basic rain barrels to more advanced systems, there is a rainwater harvesting solution for every budget and level of expertise. Educating ourselves and others about the various options can dispel the myth that rainwater harvesting is only for the wealthy or highly skilled.

Another challenge faced by proponents of rainwater harvesting is the concern about water quality. Some individuals worry that harvested rainwater may be unsafe for use, particularly for drinking. However, with proper filtration and treatment, rainwater can be as safe as other sources. Therefore, it is crucial to emphasize the importance of regular maintenance and the use of appropriate filters to ensure the safety and quality of harvested rainwater.

A further misconception is that rainwater harvesting is only suitable for regions with abundant rainfall. In reality, rainwater harvesting can be a valuable resource even in areas with limited precipitation. By implementing efficient collection and storage methods, homeowners in drier climates can still reap the benefits of rainwater harvesting. Moreover, utilizing harvested rainwater can help alleviate the strain on water resources in regions where water scarcity is a pressing issue.

To overcome these challenges and misconceptions, raising awareness and promoting education about rainwater harvesting is vital. Government agencies, industry leaders, and individuals all have a role in spreading accurate information and encouraging adoption of this sustainable practice. In addition, by offering incentives, sharing success stories, and providing resources for learning, we can foster a culture that embraces rainwater harvesting as a viable and beneficial solution for our water needs.

In conclusion, overcoming the challenges and misconceptions surrounding rainwater harvesting is crucial in promoting its widespread adoption. By dispelling myths, addressing concerns, and providing accurate information, we can pave the way for a future where rainwater harvesting is embraced as a sustainable and valuable practice for homes and communities worldwide.

The Role of Government, Industry, and Individuals in Promoting Rainwater Harvesting

As we have explored throughout this book, rainwater harvesting offers numerous benefits for our environment, communities, and households. However, to truly maximize its potential and make it a widespread practice, the collective efforts of government, industry, and individuals are crucial. Therefore, this section will delve into each stakeholder's role in promoting rainwater harvesting and fostering a sustainable future.

First and foremost, the **government** plays a pivotal role in creating a supportive framework for rainwater harvesting. By implementing policies and regulations encouraging its adoption, the government can create a favorable environment for individuals and businesses to invest in rainwater harvesting systems. This can be achieved through various means, such as offering financial incentives, tax breaks, or rebates for those who install rainwater harvesting systems. Additionally, incorporating rainwater harvesting into building codes and urban planning guidelines can promote integration into new construction projects and community developments.

On the other hand, the **industry** is responsible for innovating and improving rainwater harvesting technologies, making them more efficient, affordable, and accessible to a broader audience. Manufacturers and suppliers should develop high-quality, user-friendly systems catering to different needs and budgets. Furthermore, industry stakeholders should collaborate with educational institutions and research organizations to advance the knowledge and understanding of rainwater harvesting and train the next generation of professionals in this field.

Lastly, **individuals** play a vital role in promoting rainwater harvesting by embracing it as a sustainable practice in their daily lives. Homeowners and property owners can invest in rainwater harvesting systems, reaping the benefits of reduced water bills and a more eco-friendly lifestyle. Moreover, individuals can advocate for rainwater harvesting by sharing their experiences and knowledge with their communities, friends, and family members. This grassroots approach can help to dispel misconceptions and raise awareness about the advantages of rainwater harvesting.

In conclusion, the successful promotion of rainwater harvesting relies on the combined efforts of government, industry, and individuals. By working together, these stakeholders can create a world where rainwater harvesting is the norm, contributing to a more sustainable and water-secure future for all.

Envisioning a World Where Rainwater Harvesting is the Norm

Imagine a world where every home, office building, and public space has an efficient rainwater harvesting system. A world where the sight of rainwater tanks, catchment systems, and green roofs is as commonplace as solar panels and wind turbines. This is the world we strive for, where rainwater harvesting is not just an alternative but the norm.

In this world, we would witness a **dramatic reduction in the demand for treated water from municipal sources**, leading to decreased energy consumption and a reduced carbon footprint. In addition, our cities would be transformed into sustainable urban landscapes where green spaces and rainwater harvesting systems work harmoniously to reduce the urban heat island effect and improve air quality.

The widespread adoption of rainwater harvesting would also lead to a more resilient water supply, especially in regions prone to drought or water scarcity. In addition, by relying on a decentralized water source, communities would be better equipped to handle periods of low rainfall and reduced water availability. This would not only benefit the environment but also help alleviate the economic burden on households and businesses that often face skyrocketing water bills during water scarcity.

Moreover, embracing rainwater harvesting as the norm would foster a greater sense of environmental stewardship among individuals. As people become more aware of their water consumption and its impact on the planet, they will be more inclined to adopt other sustainable practices daily. This shift in mindset could lead to a ripple effect, inspiring further innovation and collaboration in pursuing a more sustainable future.

To achieve this vision, it is essential for rainwater harvesting to be integrated into the design and construction of new buildings and infrastructure projects. Architects, engineers, and urban planners must work together to develop innovative solutions that seamlessly incorporate rainwater harvesting systems into the fabric of our built environment. Additionally, retrofitting existing structures with rainwater harvesting technology should be encouraged and incentivized, making it accessible to more property owners.

In this envisioned world, rainwater harvesting would no longer be seen as a niche or alternative practice but as an integral component of a sustainable lifestyle. By embracing rainwater harvesting as the norm, we can collectively work towards a future where our water resources are managed responsibly, our environment is protected, and our communities are more resilient in the face of a changing climate.

Final Thoughts: The Path Forward for Rainwater Harvesting and a Sustainable Future

As we reach the end of our exploration into the world of rainwater harvesting, it is essential to take a moment to reflect on the significance of this practice in our pursuit of a sustainable future. Throughout this book, we have delved into the **numerous benefits**, **challenges**, and

opportunities that rainwater harvesting presents. This practice could revolutionize how we manage our water resources.

The path forward for rainwater harvesting is collaboration, innovation, and determination. By working together, governments, industries, and individuals can create a world where rainwater harvesting is accepted and celebrated as a vital component of sustainable living. This will require continued efforts to educate the public about the benefits of rainwater harvesting, dispel misconceptions, and encourage adopting this practice in homes and communities worldwide.

Innovation will also play a crucial role in the future of rainwater harvesting. As technology advances, we can expect new and improved systems that make rainwater harvesting more efficient and accessible. This will make it easier for individuals to incorporate rainwater harvesting into their daily lives and help drive down costs and increase the overall appeal of this practice.

As individuals, we must also recognize our role in promoting rainwater harvesting and sustainable living. By embracing this practice in our homes and sharing our experiences with others, we can create a ripple effect that will ultimately lead to the widespread adoption of rainwater harvesting. This, in turn, will have a profound impact on our environment, our communities, and our planet as a whole.

In conclusion, the journey of rainwater harvesting is far from over. As we look to the future, we must continue to champion this practice and work together to create a world where rainwater harvesting is a common practice and a cornerstone of sustainable living. By doing so, we can help to ensure that future generations inherit a planet that is healthy, vibrant, and capable of sustaining life for many years to come. So, let us embrace the potential of rainwater harvesting and take the first steps toward a more sustainable future together.

NO-TILL FARMING FOR YOUR HOME

A COMPREHENSIVE GUIDE TO SUSTAINABLE, ORGANIC CROP GROWING AND GARDENING FOR HOMEOWNERS

INTRODUCTION

There has been a growing interest in sustainable and eco-friendly gardening practices in recent years. As climate change and environmental issues become increasingly pressing, many people seek ways to reduce their ecological footprint and contribute to **a healthier planet.** One such method that has gained popularity is no-till farming. Home gardens can quickly adopt this practice to create a more sustainable and productive growing environment.

As the name suggests, no-till farming is cultivating plants without disturbing the soil through tilling or plowing. Leaving the soil undisturbed, this practice allows for preserving its natural structure, promoting a healthier ecosystem within the garden. This approach benefits the environment and leads to more **robust and bountiful harvests**, making it an attractive option for gardeners of all levels.

This book will delve into the fascinating world of no-till farming at home, exploring its history, benefits, and essential techniques. We will also discuss some common challenges faced by no-till gardeners and provide practical solutions to overcome them. By the end of this book, you will be equipped with the knowledge and tools necessary to successfully implement no-till farming practices in your home garden and contribute to a more sustainable future.

To bring you up to speed, this introduction will briefly cover the following:

- **The History and Evolution of No-Till Farming:** Discover the origins of no-till farming and how it has evolved to become the sustainable practice it is today.
- **Benefits of Adopting No-Till Practices in Your Home Garden:** Learn about the numerous advantages of no-till farming, from improved soil health to increased crop yields and reduced labor.
- **Essential Tools and Techniques for Successful No-Till Farming:** Familiarize yourself with the tools and techniques required for successful no-till gardening, including proper soil preparation, planting, and maintenance.
- **Common Challenges and Solutions in No-Till Home Gardening:** Understand the potential obstacles you may encounter in your no-till journey and find practical solutions to overcome them.
- **Embracing a Sustainable Future with No-Till Farming at Home:** Reflect on the importance of adopting sustainable practices like no-till farming and envision its positive impact on our planet and future generations.

As we embark on this journey together, you will gain valuable insights into the world of no-till farming and become an active participant in the global movement toward a more sustainable and eco-friendly future. So, let's get started and discover the incredible potential of no-till farming at home!

A Brief History of No-Till Farming

No-till farming, a practice that has gained significant attention recently, has a rich and fascinating history that dates back thousands of years. This section will delve into the origins of no-till farming, its evolution, and how it has become an essential component of sustainable agriculture today.

No-till farming can be traced back to ancient civilizations, where farmers recognized the importance of preserving soil structure and fertility. Evidence suggests that indigenous peoples in various parts of the world, including the Americas, Africa, and Asia, practiced no-till farming long before the advent of modern agriculture. These early farmers understood that disturbing the soil through plowing or tilling could lead to erosion, loss of nutrients, and decreased crop yields.

Heavy machinery and plowing became the norm as agriculture became more industrialized in the 18th and 19th centuries. The invention of the moldboard plow, which turned over the soil, burying weeds and crop residues, was considered a significant advancement in farming technology. However, this method of soil preparation also led to increased soil erosion and the depletion of soil organic matter.

The modern no-till movement began in the mid-20th century as a response to the Dust Bowl - a severe environmental disaster in the United States caused by extensive soil erosion due to poor farming practices. In the 1940s, innovative farmers and researchers started experimenting with alternative methods of soil management that would minimize soil disturbance and promote soil health. One such pioneer was Edward H. Faulkner, who published the groundbreaking book *"Plowman's Folly"* in 1943, questioning the necessity of plowing and advocating for no-till farming.

Over the next few decades, no-till farming gained momentum as researchers and farmers began recognizing its numerous benefits. In the 1960s and 1970s, the development of specialized no-till equipment, such as no-till planters and drills, made it easier for farmers to adopt this method on a larger scale. Additionally, the rise of the environmental movement during this time increased awareness of the importance of sustainable farming practices, further propelling the popularity of no-till farming.

Today, no-till farming is practiced by farmers worldwide, and its benefits are well-documented. This farming method has been shown to improve soil health, increase water retention, reduce erosion, and promote biodiversity. As a result, no-till farming has become an essential component of sustainable agriculture, helping to ensure food security and protect the environment for future generations.

The following section will explore the many benefits of adopting no-till practices in your home garden and how this approach can lead to healthier, more productive plants while contributing to a more sustainable future.

Benefits of Adopting No-Till Practices in Your Home Garden

As you embark on your journey to embrace no-till farming at home, it's essential to understand the numerous benefits this practice offers. No-till farming is not only an environmentally friendly approach to gardening, but it also provides advantages to your plants, soil, and even your wallet. In this section, we will explore the various benefits of adopting no-till practices in your home garden.

Improved Soil Health: One of the most significant advantages of no-till farming is soil health enhancement. By not tilling the soil, you preserve its natural structure and allow beneficial organisms, such as earthworms and microorganisms, to thrive. These organisms play a crucial role in breaking down organic matter and creating nutrient-rich humus, ultimately leading to the healthier, more fertile soil.

Enhanced Water Retention: No-till practices help improve the soil's ability to retain water. Since the soil structure remains intact, it can better absorb and hold moisture, reducing the need for frequent watering. This not only conserves water but also helps your plants withstand drought.

Reduced Soil Erosion: Traditional tilling methods can lead to soil erosion, as the soil's top layer is disturbed and left vulnerable to wind and water. On the other hand, no-till farming keeps the soil structure intact and covered with organic matter, reducing the risk of erosion and preserving valuable topsoil.

Increased Carbon Sequestration: No-till farming helps combat climate change by promoting carbon sequestration. When soil is left undisturbed, it can store more carbon, reducing the amount of carbon dioxide released into the atmosphere. This not only helps mitigate the effects of climate change but also contributes to overall soil fertility.

Lower Labor and Equipment Costs: No-till gardening requires less labor and equipment than traditional tilling methods. By eliminating the need for tilling, you save time and energy and reduce the wear and tear on your gardening tools. This can translate to significant cost savings over time.

Healthier Plants: Plants grown in no-till gardens are healthier and more resilient. The improved soil structure and increased presence of beneficial organisms create an optimal environment for root growth, nutrient uptake, and overall plant health. As a result, no-till gardens often produce higher yields and more robust plants.

Promotion of Biodiversity: No-till practices encourage biodiversity in your garden by providing a more natural and undisturbed environment for various organisms to coexist. This includes beneficial insects, microorganisms, birds, and other wildlife contributing to a balanced and thriving ecosystem.

In conclusion, adopting no-till practices in your home garden offers many benefits that contribute to a healthier, more sustainable, and cost-effective gardening experience. By embracing this method, you nurture your plants and soil and promote a more environmentally friendly and sustainable future.

Essential Tools and Techniques for Successful No-Till Farming

No-till farming at home is a rewarding and sustainable practice that can transform your garden into a thriving ecosystem. However, to ensure success in your no-till journey, it is crucial to familiarize

yourself with the essential tools and techniques that will make the process more efficient and enjoyable. This section will explore the must-have equipment and methods to help you achieve a flourishing no-till garden.

Critical Tools for No-Till Farming

Broadfork: A broad fork is an indispensable tool in the no-till gardener's arsenal. This large, two-handled fork is designed to gently loosen and aerate the soil without turning it over. Using a broad fork, you can maintain the soil structure and promote the growth of beneficial organisms, such as earthworms and mycorrhizal fungi.

Garden Hoe: A sturdy garden hoe is essential for breaking up compacted soil, removing weeds, and shaping planting beds. Choose a hoe with a sharp, flat blade that can easily slice through the soil and cut the roots of weeds without disturbing the soil structure.

Hand Trowel and Spade: These small, handheld tools are perfect for planting and transplanting seedlings in your no-till garden. They allow for the precise placement of plants and help minimize soil disturbance.

Mulch Materials: Organic mulch is a crucial component of no-till gardening, as it helps suppress weeds, conserve moisture, and add nutrients to the soil. Common mulch materials include straws, leaves, grass clippings, and wood chips.

Techniques for Successful No-Till Farming

Soil Preparation: Before planting, gently use a broad fork to loosen the soil and improve aeration. This will encourage root growth and promote the proliferation of beneficial soil organisms. Remove any existing weeds or grasses to prevent competition for nutrients and water.

Planting: When planting seeds or seedlings, use a hand trowel or spade to create a small hole in the soil. Gently place the plant or seed in the hole, and cover it with soil, taking care not to compact the soil around the plant. This will ensure that the roots have room to grow and access necessary nutrients.

Mulching: Apply a thick layer of organic mulch around your plants to suppress weeds, conserve moisture, and regulate soil temperature. As the mulch breaks down, it will also add valuable nutrients to the soil and improve its structure.

Crop Rotation: To maintain soil fertility and prevent the buildup of pests and diseases, practice crop rotation in your no-till garden. This involves changing the types of plants grown in a specific area each year, following a planned sequence. Crop rotation helps to balance nutrient demands, disrupt pest life cycles, and promote biodiversity.

Cover Crops: Planting cover crops, such as legumes, grasses, or other green manures, can help protect and enrich the soil during fallow periods. These plants can prevent erosion, suppress weeds, and provide a habitat for beneficial insects. Cover crops add organic matter and nutrients to the soil when terminated, further enhancing fertility.

By incorporating these essential tools and techniques into your no-till farming routine, you can create a thriving, sustainable home garden that benefits you and the environment. As you gain experience and confidence in your no-till practices, you'll discover the joy and satisfaction of nurturing a healthy, productive garden ecosystem.

Common Challenges and Solutions in No-Till Home Gardening

No-till farming at home offers numerous benefits, from improved soil health to reduced labor and expenses. However, like any gardening method, it comes with its challenges. This section will discuss some common obstacles you may encounter when implementing no-till practices in your home garden and provide practical solutions to help you overcome them.

Weed Management

One of the primary concerns for no-till gardeners is weed control. Without the traditional method of tilling to uproot and bury weeds, you may find that unwanted plants become more persistent in your garden.

Solution: To tackle weeds effectively, consider using a combination of methods, such as mulching, mowing, and hand-pulling. Mulching with organic materials like straw, wood chips, or leaves can suppress weed growth by blocking sunlight and making it difficult for weeds to take root. Mowing or cutting weeds before they go to seed can also help prevent their spread. For stubborn weeds, hand-pulling may be necessary, but remember to remove the entire root system to prevent regrowth.

Soil Compaction

In no-till gardens, soil compaction can become an issue over time, as there is no regular tilling to break up and aerate the soil. Compacted soil can hinder root growth and limit the availability of nutrients and water to your plants.

Solution: To combat soil compaction, use a broad fork or garden fork to gently loosen the soil without turning it over. This will help maintain the soil structure and promote aeration. Additionally, incorporating organic matter like compost or well-rotted manure can improve soil structure and reduce compaction.

Nutrient Availability

In a no-till system, nutrients may be less readily available to plants, particularly in the early stages of transitioning from a traditional tilled garden. This is because the decomposition of organic matter and the release of nutrients happen at a slower pace in no-till gardens.

Solution: To ensure your plants receive adequate nutrients, consider using organic fertilizers, such as compost, worm castings, or fish emulsion. These can be applied as a top dressing or mixed into the top few inches of soil. Additionally, planting cover crops and incorporating them into the soil as green manure can help boost nutrient levels and improve soil health.

Pest Management

While no-till practices can help reduce some pest problems, they can also create a favorable environment for others, such as slugs and insects that thrive in undisturbed soil.

Solution: Implement an integrated pest management (IPM)

approach, which involves using a combination of cultural, biological, and chemical controls to manage pests. This may include crop rotation, introducing beneficial insects, using barriers or traps, and applying organic pesticides when necessary. In addition, regularly monitoring your garden for signs of pests can help you address problems early and prevent them from becoming unmanageable.

Learning Curve

Transitioning to no-till gardening may require a shift in mindset and adopting new techniques, which can be challenging for those accustomed to traditional gardening methods.

Solution: Be patient and give yourself time to learn and adapt to no-till practices. Seek resources, such as books, online forums, and local gardening groups, to expand your knowledge and connect with others who share your interest in no-till gardening. Remember that every garden is unique; finding the best approach for your situation may take trial and error.

By addressing these common challenges, you can successfully implement no-till practices in your home garden and enjoy the many benefits this sustainable method offers. In the next section, we will conclude our exploration of no-till farming at home and discuss how embracing this approach can contribute to a more sustainable future for our planet.

Embracing a Sustainable Future with No-Till Farming at Home

As we reach the end of our journey exploring no-till farming at home, we must reflect on the importance of adopting this sustainable practice in our gardens. Throughout this book, we have delved into a brief history of no-till farming, discussed its numerous benefits, and provided you with the essential tools and techniques to implement this method successfully. We have also addressed common challenges home gardeners face and offered possible solutions.

Embracing no-till farming at home is more than just a gardening trend; it is a commitment to creating a healthier and more sustainable environment for ourselves and future generations. We actively participate in the global movement toward sustainable agriculture and responsible land stewardship by adopting no-till practices in our gardens.

No-till farming at home improves the health and productivity of our gardens and contributes to our planet's overall well-being. We are taking essential steps toward **combating climate change** and preserving our natural resources by reducing soil erosion, increasing water retention, and promoting biodiversity. Moreover, no-till gardening fosters a deeper connection with nature, allowing us to better understand and appreciate the intricate ecosystems within our backyards.

As you continue your no-till gardening journey, remember that patience and persistence are key. Like any new skill, mastering no-till farming takes time and practice. Be prepared to learn from your mistakes and adapt your techniques as needed. As you gain experience and confidence, you will undoubtedly see the fruits of your labor in the form of a thriving, sustainable garden.

In conclusion, no-till farming at home is an empowering and rewarding way to cultivate a healthier, more sustainable future for ourselves and our planet. By embracing this practice, we nurture our gardens and foster a deeper connection with the earth and its natural processes. As we continue to learn, grow, and adapt, we can take pride in knowing that our efforts contribute to a greener, more resilient world for future generations.

1

THE HISTORY AND EVOLUTION OF NO-TILL FARMING

A farmer using a plow to till the soil.

No-till farming, a method that has gained significant traction recently, is not a new concept. Its roots can be traced back to ancient civilizations that recognized the importance of preserving soil health and fertility. This chapter delves into the fascinating history and evolution of no-till farming, from its earliest beginnings to the modern techniques that have made it an increasingly popular choice for sustainable agriculture.

The origins of no-till farming can be traced back *thousands of years* when early agricultural societies began to understand the detrimental effects of excessive soil disturbance. These ancient farmers recognized that tilling the soil disrupted the earth's natural structure and led to soil erosion and loss of valuable nutrients. In response, they developed methods to grow crops without turning the soil, laying the foundation for what we now know as no-till farming.

As we journey through the history of no-till farming, we will explore the practices of early adopters, the impact of the agricultural revolution, and the modern innovations that have shaped this sustainable approach to agriculture. We will also examine the environmental and economic implications of no-till farming and its potential to revolutionize home gardening and small-scale farming.

By understanding the origins and evolution of no-till farming, we can appreciate the significance of this method in today's world. As we face mounting environmental challenges and a growing global population, the need for sustainable and efficient agricultural practices has never been more critical. No-till farming offers a promising solution. Its continued development and adoption hold the key to a more sustainable future for agriculture at both the large-scale and home garden levels.

Ancient Practices and Early Adopters

Though seemingly modern, the concept of no-till farming has roots that can be traced back to ancient civilizations. This section will explore the early adopters of no-till farming practices and how these methods have evolved.

Long before the advent of modern agricultural machinery, indigenous communities around the world practiced no-till farming techniques. These early adopters recognized the importance of preserving soil structure and fertility and implemented various methods to achieve this goal.

One of the earliest examples of no-till farming can be found in the ancient civilizations of the Americas. The Native Americans, particularly the tribes of the Great Plains, practiced a form of no-till agriculture known as *"slash-and-mulch."* This method involved cutting down small patches of vegetation and allowing the cut plants to decompose on the soil surface, providing a natural mulch that protected the soil from erosion and retained moisture. The Native Americans then planted their crops, such as corn, beans, and squash, directly into the mulch without disturbing the soil.

Similarly, in ancient China, farmers practiced a no-till technique called *"direct seeding,"* which involved sowing seeds directly into the soil without plowing or turning the soil. This method was prevalent in rice cultivation, as it helped to preserve soil structure and prevent erosion in the terraced fields.

In Africa, the indigenous tribes of the Congo Basin practiced a form of no-till farming called *"chitemene."* This system involved cutting down trees and shrubs, leaving them to dry for several months, and then burning the vegetation to create a layer of ash. The ash served as a natural fertilizer, and crops were planted directly into the ash without tilling the soil.

These early adopters of no-till farming understood the importance of preserving soil health and fertility, and their methods were tailored to the specific needs of their local environments. As we will see in the next section, the Agricultural Revolution brought about significant changes in farming practices, including the widespread adoption of plowing and tilling. However, the principles of no-till farming persisted, and modern innovations have allowed the development of new techniques that build upon the wisdom of these ancient practices.

The Agricultural Revolution and No-Till Farming

The Agricultural Revolution, which took place around 10,000 years ago, marked a significant turning point in human history. During this time, humans transitioned from a nomadic, hunter-gatherer lifestyle to a more settled, agricultural way of life. As communities began cultivating crops and domesticating animals, they developed various farming techniques to improve their yields and sustain their growing populations. One such technique that emerged during this period was no-till farming.

No-till farming, also known as conservation tillage or zero tillage, is a method of growing crops without disturbing the soil through plowing or tilling. This approach to agriculture has its roots in ancient practices, where farmers recognized the importance of preserving soil structure and fertility. Early adopters of no-till farming understood that minimizing soil disturbance could lead to **increased crop yields** and **healthier ecosystems.**

During the Agricultural Revolution, no-till farming was practiced in various forms across different cultures. For example, the ancient Egyptians utilized a no-till method by planting seeds directly into the fertile soil deposited by the Nile River after its annual flood. Similarly, Native American tribes, such as the Iroquois, practiced a form of no-till farming called the *"Three Sisters"* technique. This method involved planting corn, beans, and squash together in small mounds, allowing the plants to support and nourish one another without tilling the soil.

Despite the early adoption of no-till farming techniques, the practice was not widely embraced during the Agricultural Revolution. As populations grew and the demand for food increased, farmers sought more efficient ways to cultivate their land. This led to the development of plows and other tools designed to break up and turn over the soil, making it easier to plant seeds and control weeds.

Over time, these tilling practices became the norm in agriculture, and the benefits of no-till farming were largely forgotten.

It was in the *20th century* that no-till farming experienced a resurgence as modern innovations and techniques began to emerge.

Researchers and farmers alike started to recognize the negative impacts of traditional tilling practices on soil health, water quality, and the environment. This renewed interest in no-till farming led to the development of new technologies, such as specialized no-till planters and herbicides, which made it easier for farmers to adopt this sustainable approach to agriculture.

Today, no-till farming is gaining traction as a viable alternative to conventional farming methods. As we continue to learn more about the environmental and economic benefits of this practice, it is becoming increasingly clear that no-till farming can revolutionize agriculture and contribute to a more sustainable future for our planet. In the next section, we will explore the modern innovations and techniques that have made no-till farming more accessible to farmers around the world.

Modern Innovations and Techniques in No-Till Farming

In recent years, no-till farming has experienced a surge in popularity due to numerous technological advancements and innovative techniques. These modern approaches have made it easier for farmers and home gardeners alike to adopt this sustainable method of agriculture. In this section, we will explore some of the most significant innovations and techniques that have contributed to the growth and success of no-till farming in the 21st century.

One of the most notable innovations in no-till farming is the development of specialized equipment designed to **minimize soil disturbance.** No-till planters and drills, for example, have been engineered to sow seeds directly into the soil without the need for plowing or tilling. These machines have narrow, sharp discs that create a small slit in the soil, allowing the seed to be placed at the appropriate depth. This not only reduces the amount of soil disturbance but also helps to conserve moisture and prevent erosion.

Another significant advancement in no-till farming is the use of **cover crops.** Cover crops are grown between the primary crop seasons to help improve soil health, suppress weeds, and prevent erosion. In addition, they can be easily incorporated into a no-till system, as they help maintain soil structure and fertility without tilling. Some popular cover crops include clover, rye, and vetch, which can be quickly terminated with a roller-crimper or by using herbicides before planting the main crop.

Precision agriculture technology has also played a crucial role in the evolution of no-till farming.

This technology allows farmers to monitor and manage their fields more accurately and efficiently, using tools such as GPS-guided equipment, soil sensors, and drones. As a result, farmers can make more informed decisions about when and where to plant, fertilize, and irrigate their crops by collecting and analyzing data on soil conditions, nutrient levels, and crop performance. This targeted approach not only reduces the need for tillage but also helps to optimize the use of resources and minimize the environmental impact of farming.

Integrated pest management (IPM) is another modern technique that has greatly benefited no-till farming. IPM is a holistic approach to pest control that focuses on preventing and managing pests through a combination of biological, cultural, and chemical methods. By promoting natural predators, rotating crops, and using targeted, low-toxicity pesticides, farmers can effectively control pests without relying on harmful chemicals or disruptive tillage practices.

In conclusion, modern innovations and techniques in no-till farming have made it an increasingly viable and attractive option for farmers and home gardeners. By embracing these advancements, we can work towards a more sustainable and environmentally friendly future for agriculture. As we continue to develop and refine these methods, no-till farming will undoubtedly play a crucial role in shaping the future of food production and land management.

The Environmental and Economic Impact of No-Till Farming

Adopting no-till farming practices has led to significant global environmental and economic benefits for individual farmers. This section will explore the positive impacts of no-till farming and how it contributes to a more **sustainable** and **profitable** agricultural industry.

One of the most notable environmental benefits of no-till farming is its ability to reduce soil erosion. Traditional tilling methods can strip the soil of its protective layer of plant residue, leaving it vulnerable to erosion by wind and water. On the other hand, no-till farming maintains this protective layer, helping preserve the integrity of the soil and prevent the loss of valuable topsoil. This, in turn, helps to maintain soil fertility and productivity, ensuring a healthier and more sustainable agricultural system.

Another critical environmental advantage of no-till farming is its potential to improve water quality. By reducing soil erosion, no-till farming practices can also decrease the amount of sediment and agricultural chemicals that enter waterways. This helps protect aquatic ecosystems and reduce the risk of harmful algal blooms, which can devastate wildlife and human health.

No-till farming also plays a crucial role in mitigating climate change by increasing the soil's capacity to store carbon. When soil is tilled, the carbon stored within it is released into the atmosphere as carbon dioxide, a potent greenhouse gas. By leaving the soil undisturbed, no-till farming allows for more carbon sequestration, reducing agriculture's overall carbon footprint and helping to combat global warming.

From an economic standpoint, no-till farming offers numerous benefits to farmers. One of the most immediate advantages is reduced labor and fuel costs associated with traditional tilling practices. By eliminating the need for tilling, farmers can save both **time** and **money,** allowing them to allocate resources more efficiently.

Also, no-till farming can increase crop yields, as healthier soil promotes stronger root systems and improved nutrient uptake. This, in turn, can result in higher profits for farmers, as they can produce more crops with fewer inputs. Furthermore, reducing soil erosion can help preserve the long-term productivity of farmland, ensuring a more stable and secure income for farmers in the future.

In conclusion, the environmental and economic impacts of no-till farming are overwhelmingly positive. By reducing soil erosion, improving water quality, and mitigating climate change, no-till farming practices contribute to a more sustainable and resilient agricultural system. At the same time, these practices can lead to increased profits for farmers, making no-till farming an attractive and viable option for the future of agriculture. As we continue to face the challenges of a growing global population and a changing climate, adopting no-till farming at home and on a larger scale will become increasingly important in ensuring a sustainable and prosperous future for all.

The Future of No-Till Farming at Home

As we have journeyed through the history and evolution of no-till farming, it is evident that this innovative approach to agriculture has come a long way. From its ancient roots to the modern techniques employed today, no-till farming has proven to be a sustainable and environmentally friendly method of growing crops. Therefore, as we look towards the future, we must consider how no-till farming can be further integrated into home gardening practices, ultimately contributing to a greener and more sustainable world.

The growing awareness of the environmental and economic benefits of no-till farming has led to a surge in its adoption by both large-scale farmers and home gardeners. Furthermore, with the increasing availability of resources and tools tailored to the needs of small-scale no-till farming, it is becoming more accessible for individuals to implement these practices in their backyards. This shift towards sustainable home gardening can significantly reduce the environmental impact of our food systems and promote a deeper connection to the land and the food we consume.

In the future, we expect to see continued advancements in no-till farming techniques and technologies. These innovations will likely make it even easier for home gardeners to adopt no-till practices, further reducing the barriers to entry. Additionally, as more people become aware of the benefits of no-till farming, communities will likely come together to share knowledge, resources, and support, fostering a culture of sustainable agriculture at the grassroots level.

Moreover, the future of no-till farming at home will likely involve a greater emphasis on regenerative agriculture, which focuses on improving soil health, increasing biodiversity, and enhancing ecosystem resilience. Home gardeners can play a vital role in mitigating climate change and preserving the planet for future generations by adopting regenerative practices in conjunction with no-till farming.

In conclusion, the future of no-till farming at home is bright and full of potential. As more individuals embrace this sustainable approach to gardening, we can expect to see a positive shift in the way we grow and consume food. Furthermore, by continuing to innovate and share knowledge, we can work together to create a healthier, more sustainable world for ourselves and future generations. So, let us roll up our sleeves, grab our no-till tools, and embark on this exciting journey toward a greener and more bountiful future.

Chapter Summary

1. No-till farming has ancient roots, with early agricultural societies recognizing the importance of preserving soil health and fertility by minimizing soil disturbance.
2. Early adopters of no-till farming included Native Americans, ancient Chinese farmers, and indigenous tribes of the Congo Basin, who developed methods tailored to their local environments.
3. The Agricultural Revolution led to the development of plows and other tools for soil cultivation, causing no-till farming to be largely forgotten until the 20th century.
4. Modern innovations in no-till farming include specialized equipment, cover crops, precision agriculture technology, and integrated pest management, making it more accessible for farmers and home gardeners.
5. No-till farming offers significant environmental benefits, such as reduced soil erosion, improved water quality, and

increased carbon sequestration, which helps mitigate climate change.

6. Economically, no-till farming can reduce labor and fuel costs, increase crop yields, and long-term preservation of farmland productivity, resulting in higher farmer profits.
7. The future of no-till farming at home involves greater integration of sustainable and regenerative agriculture practices, fostering a culture of environmentally friendly gardening.
8. As more individuals adopt no-till farming and regenerative practices, we can expect a positive shift in how we grow and consume food, contributing to a healthier and more sustainable world.

2

UNDERSTANDING SOIL HEALTH AND ITS IMPORTANCE

A plant growing in soil, with roots reaching deep into the ground and interacting with the soil ecosystem to absorb nutrients and water.

Imagine a bustling city teeming with life and activity, where each individual plays a vital role in maintaining the harmony and balance of the community. Now, picture this city beneath your feet, hidden in the ground you walk on. This is the fascinating world of soil and the foundation of a successful no-till home garden.

Soil health is a crucial aspect of no-till farming, as it directly influences the growth and productivity of your plants. In this chapter, we will explore the importance of soil health in no-till farming, delve into the components that make up healthy soil, and discuss the benefits of improved soil health in home gardens. We will also provide practical techniques for enhancing soil health in your no-till garden and guide you on how to monitor and assess the health of your soil.

At its core, soil health refers to the capacity of soil to function as a living ecosystem, supporting plant and animal life, maintaining environmental quality, and promoting overall plant health. No-till farming focuses on preserving and enhancing this natural ecosystem rather than disrupting it through traditional tilling practices. By understanding the role of soil health in no-till farming, you can create a thriving, sustainable garden that produces bountiful harvests and contributes to a healthier environment.

In the following sections, we will uncover the secrets of healthy soil and provide you with the knowledge and tools necessary to cultivate a flourishing no-till home garden. So, let's embark on this exciting journey together and discover the wonders of soil health and its role in no-till farming.

The Components of Healthy Soil: Physical, Chemical, and Biological Properties

This section will delve into the three essential aspects of healthy soil: physical, chemical, and biological properties. Understanding these components is crucial for maintaining and improving soil health in your no-till home garden.

Physical Properties

The physical properties of soil refer to its texture, structure, and porosity. Soil texture is determined by the relative proportions of sand, silt, and clay particles. A well-balanced soil, known as loam, contains a mixture of all three particle sizes, providing an ideal environment for plant growth. Soil structure refers to the arrangement of these particles and the spaces between them. Good soil structure allows proper water infiltration, aeration, and root penetration.

Porosity, or the amount of pore space within the soil, is another important physical property. Pore spaces are critical for holding water, air, and nutrients vital for plant growth. In no-till farming, practices such as cover cropping and avoiding soil compaction help maintain and improve soil porosity.

Chemical Properties

The chemical properties of soil include its pH, nutrient content, and cation exchange Capacity (CEC). Soil pH is a measure of acidity or alkalinity, which affects nutrient availability and microbial activity. Most plants thrive in slightly acidic to neutral soils (pH 6-7), while some may prefer slightly alkaline conditions. Regular soil testing can help you monitor and adjust your soil's pH to optimize plant growth.

Nutrient content refers to the presence of essential elements such as nitrogen, phosphorus, and potassium, which are vital for plant growth and development. Incorporating organic matter and using compost are no-till farming practices that can help maintain and improve nutrient levels in the soil.

Cation exchange capacity (CEC) is a measure of the soil's ability to hold and exchange positively charged ions (cations) such as calcium, magnesium, and potassium. A higher CEC indicates a greater capacity to retain nutrients and resist changes in pH. Organic matter and clay particles contribute to a soil's CEC, making it an essential factor to consider in no-till farming.

Biological Properties

The biological properties of soil encompass the diverse community of microorganisms, such as bacteria, fungi, and protozoa, as well as larger organisms like earthworms and insects. These organisms play a crucial role in maintaining soil health by breaking down organic matter, fixing nitrogen, and enhancing soil structure.

Using cover crops, applying compost, and minimizing soil disturbance are no-till farming practices that help promote a thriving soil ecosystem. A healthy soil food web benefits plant growth and contributes to natural pest control and disease suppression.

In conclusion, understanding a healthy soil's physical, chemical, and biological properties is essential for successful no-till farming at home. By implementing practices that support these components, you can create a fertile and sustainable environment for your plants to thrive.

The Benefits of Improved Soil Health in No-Till Home Gardens

Healthy soil is the foundation of any successful garden, especially for no-till home gardens. By understanding and prioritizing soil health, you can unlock a myriad of benefits that will not only improve your garden's productivity but also contribute to a more sustainable and eco-friendly approach to gardening. This section will explore the numerous advantages of improved soil health in no-till home gardens.

First and foremost, healthy soil promotes vigorous plant growth. When your soil's physical, chemical, and biological properties are balanced, your plants can access the essential nutrients, water, and air they need to thrive. This results in stronger root systems, more robust foliage, and a bountiful harvest of fruits, vegetables, and flowers.

Another significant benefit of improved soil health is increased resistance to pests and diseases. A well-balanced soil ecosystem encourages the presence of beneficial microorganisms and insects that help to keep harmful pests and pathogens in check. This natural form of pest control reduces the need for synthetic pesticides, which can harm both the environment and the health of your family.

Improved soil health also leads to better water retention and infiltration. Healthy soil has a more porous structure, which allows water to penetrate more easily and be stored for longer periods. This means your plants will have a more consistent water supply, even during dry spells. Additionally, better water infiltration reduces the risk of runoff and erosion, protecting your precious topsoil and preventing pollutants from entering nearby water sources.

One of the most notable benefits of healthy soil in no-till home gardens is increased fertility and nutrient availability. As organic matter decomposes, it releases essential nutrients that plants need to grow. In a no-till system, these nutrients are more readily available to plant roots, as they are not lost through tillage-induced erosion or leaching. This means that you may need to apply fewer synthetic fertilizers, saving you money and reducing the environmental impact of your garden.

Lastly, improved soil health contributes to the overall sustainability of your garden. By fostering a diverse and thriving soil ecosystem, you promote a more resilient and self-sufficient garden requiring fewer external inputs. This benefits your immediate surroundings and contributes to the global effort to reduce our reliance on non-renewable resources and mitigate the impacts of climate change.

In conclusion, the benefits of improved soil health in no-till home gardens are numerous and far-reaching. By prioritizing the well-being of your soil, you can enjoy a more productive, resilient, and sustainable garden that provides a bounty of fresh produce and beautiful blooms for you and your family to enjoy.

Techniques for Enhancing Soil Health in No-Till Farming

In the world of no-till farming, soil health is paramount. A vibrant, nutrient-rich soil is the foundation for a thriving home garden. For this to be achieved, it's essential to employ various techniques that enhance and maintain soil health. In this section, we will explore some of the most effective methods for nurturing your soil and ensuring the success of your no-till garden.

Cover Crops: One of the most effective ways to improve soil health is by planting cover crops. These plants, such as clover, rye, or vetch, help prevent soil erosion, suppress weeds, and add organic matter to the soil. When cover crops are terminated, their decomposing roots and shoots provide valuable nutrients to the soil, improving its fertility and structure.

Organic Matter Amendments: Adding organic matter, such as compost or well-rotted manure, to your soil is another excellent way to enhance health. Organic matter improves soil structure, increases water-holding capacity, and provides essential nutrients for plant growth. Additionally, it supports the growth of beneficial microorganisms, which play a crucial role in nutrient cycling and disease suppression.

Mulching: Applying a layer of organic mulch, such as straw, wood chips, or leaves, on the soil surface is an effective technique for conserving soil moisture, suppressing weeds, and moderating soil temperature. Mulch also breaks down over time, adding valuable organic matter to the soil and promoting the growth of beneficial soil organisms.

Crop Rotation: Rotating the types of crops you grow in your no-till garden can help maintain soil health by preventing the build-up of pests and diseases and ensuring a balanced nutrient supply. For example, alternating between nitrogen-fixing legumes and nutrient-demanding crops like tomatoes or peppers can help maintain soil fertility and reduce the need for additional fertilizers.

Plant Diversity: Including a variety of plants in your no-till garden can contribute to improved soil health. Different plant species have unique root structures and nutrient requirements, which can help maintain a balanced soil ecosystem. Plant diversity also supports a broader range of beneficial insects and microorganisms, which can help suppress pests and diseases.

Minimizing Soil Disturbance: In no-till farming, it's essential to minimize soil disturbance to preserve its structure and the delicate balance of microorganisms within it. Avoid walking on planting beds, and use tools like broad forks or garden forks to gently loosen the soil without turning it over. This approach helps maintain soil structure and promotes the growth of beneficial organisms.

By implementing these techniques in your no-till home garden, you can enhance and maintain the health of your soil, ensuring a productive and sustainable growing environment for your plants. Remember, healthy soil is the foundation of a successful no-till garden. With little care and attention, you can cultivate a thriving ecosystem that will reward you with bountiful harvests for years.

Monitoring and Assessing Soil Health in Your Home Garden

To ensure the success of your no-till home garden, monitoring and assessing your soil's health regularly is crucial. By closely monitoring your garden's soil, you can identify any potential issues early on and take the necessary steps to maintain or improve its health. In this section, we will discuss various methods and tools that can help you effectively monitor and assess the soil health in your home garden.

One of the simplest ways to monitor soil health is through visual observation. As you tend to your garden, take note of the soil's color, texture, and structure. Healthy soil should be dark in color, crumbly in texture, and have a well-defined structure. Additionally, observe the presence of earthworms and other soil organisms, as they are indicators of good soil health.

Another essential aspect of monitoring soil health is testing its chemical properties. Home gardeners can purchase soil testing kits from local garden centers or online retailers. These kits typically measure the soil's pH, nitrogen, phosphorus, and potassium levels. By understanding the nutrient content of your soil, you can make informed decisions about the need for amendments or fertilizers.

To assess the biological properties of your soil, you can perform a simple DIY test called the "soil food web" assessment. This involves taking a small soil sample and observing it under a microscope or magnifying glass. Look for the presence of microorganisms such as bacteria, fungi, and protozoa, which are essential for breaking down organic matter and releasing nutrients for plant uptake.

In addition to these methods, various tools and technologies are available to help monitor soil health. For example, soil moisture meters can help determine if your garden receives the right amount of water. At the same time, penetrometers can measure soil compaction levels. Investing in these tools can provide valuable insights into the health of your soil and guide your no-till gardening practices.

As you monitor and assess your soil's health, remember that changes may not occur overnight. Improving soil health is a gradual process that requires patience and persistence. Keep track of your observations and test results in a garden journal, and adjust your no-till practices as needed.

In conclusion, monitoring and assessing soil health is vital to successful no-till home gardening. By using a combination of visual observation, testing, and specialized tools, you can understand your soil's health comprehensively and make informed decisions to ensure the long-term success of your no-till garden.

The Long-Term Impact of Soil Health on No-Till Farming Success

In conclusion, the long-term success of no-till farming at home is intrinsically linked to soil health. As we have explored throughout this chapter, understanding and maintaining soil health is crucial for your home garden's overall productivity and sustainability. By focusing on the soil's physical, chemical, and biological properties, you can create a thriving environment for your plants to grow and flourish.

The benefits of improved soil health in no-till home gardens are numerous, from increased nutrient availability and water retention to reduced erosion and enhanced biodiversity. These advantages contribute to the growth of healthier, more resilient plants and promote a more sustainable and eco-friendly approach to gardening.

By implementing the various techniques discussed in this chapter, such as incorporating organic matter, practicing crop rotation, and minimizing soil disturbance, you can actively enhance the health of your soil. These practices work together to create a balanced ecosystem within your garden, providing the ideal conditions for your plants to thrive.

Monitoring and assessing soil health is an ongoing process, and staying informed about the state of your garden's soil is essential. Regular soil testing and observation will help you make informed decisions about the necessary amendments and adjustments to maintain optimal soil health.

In the long run, focusing on soil health in your no-till home garden will lead to increased productivity, reduced input costs, and a more sustainable approach to gardening. By nurturing the complex ecosystem beneath our feet, we can ensure the long-term success of no-till farming and contribute to a healthier, more vibrant world for future generations. So, let us embrace the power of healthy soil and embark on this exciting journey towards a more sustainable and fruitful home garden.

Chapter Summary

1. Soil health is crucial for the success of no-till farming, as it directly influences plant growth and productivity. Understanding the role of soil health in no-till farming can lead to a thriving, sustainable garden.
2. A healthy soil consists of balanced physical, chemical, and biological properties. These include soil texture, structure, porosity, pH, nutrient content, cation exchange capacity, and the presence of diverse microorganisms and larger organisms.
3. Improved soil health in no-till home gardens offers numerous benefits, such as vigorous plant growth, increased resistance to pests and diseases, better water retention and infiltration, increased fertility and nutrient availability, and overall sustainability.

4. Techniques for enhancing soil health in no-till farming include planting cover crops, adding organic matter amendments, mulching, practicing crop rotation, maintaining plant diversity, and minimizing soil disturbance.
5. Monitoring and assessing soil health is essential for successful no-till home gardening. Visual observation, soil testing, and specialized tools can help gardeners understand their soil's health and make informed decisions.
6. Regular soil testing and observation can help identify potential issues early on and guide necessary amendments or adjustments to maintain optimal soil health.
7. Improving soil health is a gradual process that requires patience and persistence. However, keeping track of observations and test results in a garden journal can help gardeners adjust their no-till practices as needed.
8. Focusing on soil health in no-till home gardens leads to increased productivity, reduced input costs, and a more sustainable approach to gardening, ultimately contributing to a healthier, more vibrant world for future generations.

3

THE SCIENCE BEHIND NO-TILL FARMING TECHNIQUES

A towering skyscraper, but instead of windows, it's covered in a thick blanket of wild weeds growing out of control.

No-till farming, a revolutionary approach to agriculture, has gained significant attention in recent years due to its numerous environmental and economic benefits. At its core, no-till farming is a method that eliminates the need for plowing or tilling the soil, allowing for the natural processes of the earth to take center stage. This chapter delves into the fascinating science behind no-till farming techniques, exploring how this innovative approach contributes to a more sustainable and productive agricultural landscape.

Traditional farming practices often involve turning over the soil to prepare it for planting, which can lead to soil erosion, loss of organic matter, and disruption of the delicate soil ecosystem. In contrast, no-till farming techniques prioritize soil health, recognizing that a thriving soil ecosystem is a foundation for successful agriculture. In addition, by leaving the soil undisturbed, no-till farming promotes a diverse and active community of microorganisms, enhancing soil fertility and structure.

The science of no-till farming is rooted in the understanding that soil is a living, breathing entity teeming with life. From microscopic bacteria and fungi to larger organisms like earthworms, the soil is home to countless creatures that play a crucial role in maintaining a healthy and productive environment for plants to grow. By fostering this intricate web of life, no-till farming techniques support the natural processes that drive nutrient cycling, carbon sequestration, and pest and weed control.

In the following sections, we will explore the various aspects of the science behind no-till farming, including the role of soil health and microbial activity, the impact on carbon sequestration and climate change mitigation, and the strategies for nutrient cycling and fertility management. We will also discuss the innovative pest and weed control methods that are integral to the success of no-till farming systems.

As we journey through the science of no-till farming, it is essential to remember that this approach is not just about the technical aspects of agriculture. It is a holistic mindset that embraces the interconnectedness of the natural world, recognizing that the health of our soil, plants, and ecosystems are all intertwined. By understanding and implementing the principles of no-till farming, we can work towards a more sustainable and resilient future for our planet and its inhabitants.

Soil Health and Microbial Activity in No-Till Systems

This section delves into the fascinating world beneath our feet, exploring the intricate relationship between soil health and microbial activity in no-till farming systems. As we uncover the science behind this sustainable agricultural technique, we will see how no-till farming preserves soil structure and fosters a thriving ecosystem of beneficial microorganisms.

To understand the significance of soil health, we must first recognize that soil is a living, breathing entity. It is teeming with billions of microorganisms, including bacteria, fungi, protozoa, and nematodes. These microscopic creatures are crucial in maintaining soil fertility, breaking down organic matter, and promoting plant growth. The soil's natural processes are preserved in no-till systems, allowing these microorganisms to flourish and contribute to a healthy, productive garden.

One of the primary benefits of no-till farming is the **preservation of soil structure.** When soil is left undisturbed, it retains its natural layers and aggregates, which are essential for water infiltration, aeration, and root development. This stable environment allows microorganisms to establish complex networks and symbiotic relationships with plant roots. For instance, mycorrhizal fungi form a mutually beneficial partnership with plant roots, exchanging nutrients and enhancing the plant's ability to absorb water and nutrients from the soil.

No-till farming also promotes the growth of beneficial bacteria, which are essential for breaking down organic matter and releasing nutrients into the soil. In a no-till system, organic matter is left on the soil surface, providing a constant food source for these bacteria. In addition, as they decompose the organic matter, they release nutrients such as nitrogen, phosphorus, and potassium, which are vital for plant growth. This process, known as mineralization, is a crucial component of nutrient cycling in no-till systems.

Another critical aspect of soil health in no-till farming is the role of earthworms and other soil-dwelling organisms. These creatures help to aerate the soil, improve water infiltration, and break down organic matter, further contributing to a healthy soil ecosystem. In no-till systems, the absence of tillage allows earthworm populations to thrive, enhancing the overall soil structure and fertility.

In conclusion, the science behind no-till farming techniques reveals a fascinating interplay between soil health and microbial activity. By preserving soil structure and fostering a thriving ecosystem of beneficial microorganisms, no-till farming promotes a sustainable and productive environment for growing plants. As we continue to explore the science of no-till farming, we will see how this approach benefits the soil and contributes to climate change mitigation, nutrient cycling, and effective pest and weed control strategies.

Carbon Sequestration and Climate Change Mitigation

In today's world, climate change is a pressing issue that demands immediate attention and action. Carbon sequestration is one of the most significant ways no-till farming contributes to mitigating climate change. This section will explore the science behind carbon sequestration and how no-till farming techniques play a crucial role in combating climate change.

Carbon sequestration refers to the process of capturing and storing atmospheric carbon dioxide (CO_2) in various forms, such as plant biomass and soil organic matter. This process helps reduce the amount of CO_2 in the atmosphere, thereby mitigating the greenhouse effect and slowing down global warming. No-till farming techniques are particularly effective at promoting carbon sequestration due to the minimal soil disturbance and preservation of organic matter.

In conventional farming practices, tilling the soil exposes it to oxygen, which leads to the rapid decomposition of organic matter by soil microbes.

This process releases CO_2 back into the atmosphere, contributing to the greenhouse effect. On the other hand, no-till farming leaves the soil undisturbed, allowing organic matter to accumulate and store carbon. This not only **reduces CO_2 emissions** but also **improves soil health** and **fertility**.

Moreover, no-till farming promotes the growth of plant roots, which play a significant role in carbon sequestration. As plants grow, they absorb CO_2 from the atmosphere through photosynthesis and convert it into carbohydrates. These carbohydrates are then transported to the roots, where they are either stored as root biomass or exuded into the soil as root exudates. These exudates serve as a food source for soil microbes, which in turn produce humus – a stable form of organic matter that can store carbon for hundreds or even thousands of years.

Another essential aspect of carbon sequestration in no-till farming systems is the role of cover crops. Cover crops are plants grown between the main cash crops to protect and improve the soil. They help sequester carbon by adding organic matter to the soil, reducing erosion, and promoting the growth of soil microbes. When cover crops are used in conjunction with no-till farming techniques, the potential for carbon sequestration is significantly enhanced.

In conclusion, no-till farming techniques offer a powerful tool in the fight against climate change by promoting carbon sequestration in the soil. By minimizing soil disturbance, encouraging the growth of plant roots, and utilizing cover crops, no-till farming helps to capture and store atmospheric CO_2, mitigating the greenhouse effect and contributing to a more sustainable future. As we continue to face the challenges of climate change, embracing the science behind no-till farming is essential for creating a healthier, more resilient environment for future generations.

Nutrient Cycling and Fertility Management

This section will delve into the fascinating world of nutrient cycling and fertility management in no-till farming systems. Understanding the science behind these processes can optimize your home garden's productivity while maintaining a healthy and sustainable ecosystem.

Nutrient cycling is the process by which essential elements, such as nitrogen, phosphorus, and potassium, are transformed and moved through the soil, plants, and atmosphere. The natural decomposition and nutrient release processes are enhanced in a no-till system, leading to improved soil fertility and plant health.

One of the key players in nutrient cycling is the *soil food web*, a complex network of microorganisms, including bacteria, fungi, and other tiny creatures that break down organic matter and release nutrients back into the soil. The soil food web remains intact and undisturbed in a no-till garden, allowing these microorganisms to thrive and efficiently recycle nutrients.

Fertility management in no-till systems relies on the addition of organic matter, such as compost, cover crops, and mulches, to provide a continuous source of nutrients for the soil food web. These materials supply essential nutrients and improve soil structure and water-holding capacity, further enhancing the garden's overall health and productivity.

The nitrogen cycle is one of the most important aspects of fertility management in no-till farming. Nitrogen is a critical nutrient for plant growth, and the activity of nitrogen-fixing bacteria largely determines its availability in the soil. These bacteria convert atmospheric nitrogen into a form that plants can use, known as ammonium. In a no-till system, the presence of cover crops, such as legumes, can provide a natural source of nitrogen fixation, reducing the need for synthetic fertilizers.

Phosphorus and potassium are also vital nutrients for plant growth, and the activity of the soil food web influences their availability in the soil. In no-till systems, the addition of organic matter and the maintenance of a healthy soil food web help to ensure that these nutrients are readily available for plant uptake.

Pest and weed control strategies in no-till farming systems also contribute to nutrient cycling and fertility management. In addition, minimizing soil disturbance and maintaining a diverse plant community allows no-till gardeners to create a balanced ecosystem that naturally suppresses pests and weeds, reducing the need for chemical interventions.

In conclusion, the science behind nutrient cycling and fertility management in no-till farming systems offers a compelling argument for adopting these practices in your home garden. By working with nature and harnessing the power of the soil food web, you can create a thriving, sustainable garden that provides abundant, nutrient-rich produce for your family.

Pest and Weed Control Strategies in No-Till Farming

This section will explore the fascinating world of pest and weed control strategies employed in no-till farming systems. The beauty of these methods lies in their ability to maintain a healthy and balanced ecosystem while effectively managing unwanted intruders. By understanding the science behind these strategies, you can implement them in your no-till garden and enjoy the benefits of a thriving, sustainable environment.

One of the key principles of no-till farming is **working with nature** rather than against it. Instead of relying on synthetic chemicals to combat pests and weeds, no-till farmers utilize natural processes and biological controls to keep these nuisances in check. Let's explore some of these innovative techniques.

Crop Rotation and Diversity

Crop rotation is a time-tested method of disrupting the life cycles of pests and weeds. By changing the types of plants grown in a particular area each season, no-till farmers can prevent the buildup of pests and diseases that are specific to certain crops. Additionally, planting diverse crops can attract beneficial insects that prey on pests, reducing the need for chemical interventions.

Cover Crops and Mulching

Cover crops and mulching play a crucial role in no-till pest and weed management. Cover crops, such as clover or rye, can be grown between main crops to suppress weeds and provide a habitat for beneficial insects. When these cover crops are terminated, they form a thick layer of mulch that not only smothers weeds but also creates a physical barrier that makes it difficult for pests to reach the soil and lay eggs.

Biological Control Agents

No-till farming encourages the presence of natural predators, such as ladybugs, lacewings, and predatory mites, which help to keep pest populations in check. By providing a diverse and healthy ecosystem, no-till farmers can support these beneficial insects and reduce the need for chemical pesticides. Additionally, some no-till farmers may introduce specific biological control agents, such as nematodes or bacteria, to target specific pests.

Mechanical and Manual Control

In some cases, no-till farmers may need to employ mechanical or manual control methods to manage pests and weeds. This can include hand-weeding, hoeing, or using specialized equipment to remove weeds without disturbing the soil structure. These methods are generally less disruptive to the environment than chemical controls. They can be an effective component of an integrated pest management strategy.

Integrated Pest Management (IPM)

No-till farming is a perfect fit for the principles of integrated pest management (IPM), which emphasizes the use of multiple control methods to manage pests and weeds in an environmentally friendly manner. By combining crop rotation, cover crops, biological control agents, and mechanical or manual control methods, no-till farmers can create a robust and resilient system that minimizes the need for chemical interventions.

In conclusion, the science behind pest and weed control strategies in no-till farming is both fascinating and effective. By working with nature and employing a combination of techniques, no-till farmers can create a balanced ecosystem that promotes healthy plant growth while keeping pests and weeds at bay. As you embrace these strategies in your no-till garden, you'll be contributing to a more sustainable and harmonious future for our planet.

Embracing the Science for a Sustainable Future

As we conclude our exploration of the science behind no-till farming techniques, we must recognize the far-reaching implications of adopting these practices in our home gardens and larger agricultural systems. By understanding the complex relationships between soil health, microbial activity, carbon sequestration, nutrient cycling, and pest and weed control, we can make informed decisions that promote sustainability and resilience in our food production systems.

No-till farming is more than just a method of growing plants; it is a holistic approach to agriculture that prioritizes the long-term health of our soil, ecosystems, and planet. By fostering a thriving underground community of microorganisms, we can enhance the fertility and productivity of our gardens without relying on synthetic fertilizers and harmful chemicals. This not only benefits our plants and the food we consume but also contributes to the overall well-being of our environment.

Moreover, the role of no-till farming in mitigating climate change cannot be overstated. By sequestering carbon in the soil and reducing greenhouse gas emissions, we can play an active part in addressing one of the most pressing challenges of our time. As responsible stewards of the Earth, we have to adopt practices that minimize our ecological footprint and promote the health of our planet for future generations.

Furthermore, no-till farming techniques offer practical solutions for managing pests and weeds without resorting to harmful pesticides and herbicides. By fostering a balanced ecosystem and encouraging natural predators, we can maintain the health of our gardens while protecting the biodiversity essential for a thriving environment.

In conclusion, embracing the science of no-till farming techniques is not only an investment in the health of our gardens but also a commitment to a sustainable future. By understanding and applying these principles, we can cultivate a deeper connection with the natural world, contribute to the fight against climate change, and ensure the long-term viability of our food systems. So, as we continue to learn and grow, let us remember that the choices we make in our gardens today have the power to shape the world of tomorrow.

Chapter Summary

1. No-till farming is a sustainable agricultural approach that eliminates the need for plowing or tilling the soil, promoting soil health and a thriving ecosystem of microorganisms.
2. Soil health and microbial activity are crucial components of no-till farming, as they enhance soil fertility and structure, allowing for better nutrient cycling and plant growth.
3. No-till farming contributes to climate change mitigation through carbon sequestration, capturing and storing atmospheric carbon dioxide in plant biomass and soil organic matter.
4. Nutrient cycling and fertility management in no-till systems rely on the addition of organic matter and the maintenance of a healthy soil food web, reducing the need for synthetic fertilizers.
5. Pest and weed control strategies in no-till farming focus on working with nature, utilizing crop rotation, cover crops, biological control agents, and mechanical or manual control methods.
6. Integrated Pest Management (IPM) principles align well with no-till farming, emphasizing the use of multiple

environmentally friendly control methods to manage pests and weeds.

7. Adopting no-till farming practices in home gardens and larger agricultural systems promotes long-term soil health and ecosystem balance and contributes to the fight against climate change.
8. Embracing the science of no-till farming techniques is an investment in a sustainable future, fostering a deeper connection with the natural world and ensuring the long-term viability of our food systems.

4

PREPARING YOUR HOME GARDEN FOR NO-TILL FARMING

A wise old farmer sitting on a rocking chair, overlooking his vast farmland as he shares his farming philosophy with his grandchild.

As you embark on your journey to create a thriving no-till home garden, it's essential to first embrace the philosophy behind this sustainable and eco-friendly approach to gardening. No-till farming is more than just a method; it's a mindset that prioritizes the long-term health of your soil and the environment over short-term convenience. By understanding and adopting the principles of no-till gardening, you'll be well on your way to cultivating a flourishing, productive garden that benefits both you and the planet.

At the core of the no-till philosophy is the belief that soil is a living, breathing ecosystem teeming with life. This complex network of microorganisms, fungi, and other organisms plays a crucial role in maintaining soil health and fertility. Traditional tilling methods, which involve turning over the soil to prepare it for planting, can disrupt this delicate balance and lead to soil degradation, erosion, and loss of valuable nutrients. On the other hand, no-till gardening seeks to preserve and enhance the soil's natural structure and vitality by minimizing disturbance and promoting organic matter accumulation.

To fully embrace the no-till philosophy, it's important to shift your perspective on gardening from one of control and domination to one of stewardship and cooperation. Instead of forcing your garden to conform to your desires, you'll learn to work in harmony with nature, nurturing the soil and its inhabitants to create a resilient, self-sustaining ecosystem. This approach results in healthier, more productive plants and fosters a deeper connection to the earth and a greater appreciation for the intricate web of life that sustains us all.

As you prepare your home garden for no-till farming, remember that this is a long-term investment in the health of your soil and the environment. It may take time for your garden to adapt to this new approach fully. Still, with patience, persistence, and a commitment to the no-till philosophy, you'll soon see the fruits of your labor in the form of a **thriving, vibrant garden** that nourishes both body and soul.

Assessing Your Soil

Before diving into the world of no-till gardening, it's essential to understand the importance of healthy soil. Soil is the foundation of any garden, and its quality directly impacts the success of your no-till endeavor. This section will explore how to assess your soil and make any necessary improvements to create a thriving, sustainable, no-till garden.

Know Your Soil Type

The first step in assessing your soil is to determine its type. There are three primary soil types: **sandy, loamy, and clay.** Each type has its unique characteristics, and understanding these differences will help you make informed decisions about plant selection and soil amendments.

To identify your soil type, simply grab a handful of moist soil and give it a squeeze. If it crumbles easily and doesn't hold its shape, you have **sandy soil.** If it forms a ball but breaks apart when poked, you have **loamy soil.** And if it forms a hard, compact ball that doesn't break apart easily, you have **clay soil.**

Test Your Soil's pH Level

The pH level of your soil is a crucial factor in determining the availability of nutrients for your plants. Most plants thrive in slightly acidic to neutral soil, with a pH between 6.0 and 7.0. To test your soil's pH, you can purchase a soil testing kit from your local garden center or send a sample to a soil testing laboratory.

If your soil's pH is too high (alkaline) or too low (acidic), you can amend it with organic materials. For example, adding sulfur or peat moss can lower the pH, while adding lime can raise it. Be sure to follow the recommendations provided by your soil test results to ensure proper pH balance.

Evaluate Soil Drainage

Proper drainage is essential for a successful no-till garden, as it prevents waterlogged soil and encourages healthy root growth. To test your soil's drainage, dig a hole about 12 inches deep and fill it with water. Allow the water to drain completely, then fill the hole again. Suppose the water takes longer than four hours to drain. In that case, you may need to improve your soil's drainage by incorporating organic matter, such as compost or aged manure.

Assess Soil Fertility

Healthy, fertile soil is rich in organic matter and nutrients, vital for plant growth. To assess your soil's fertility, examine its color and texture. Dark, crumbly soil is a good indicator of high organic matter content. Additionally, consider conducting a soil test to determine the levels of essential nutrients, such as nitrogen, phosphorus, and potassium. Based on the test results, you can amend your soil with organic fertilizers or other nutrient-rich materials to improve fertility.

Observe Soil Life

A thriving no-till garden relies on a diverse and active community of soil organisms, such as earthworms, bacteria, and fungi. These organisms help break down organic matter, improve soil structure, and make nutrients available to plants. To assess the health of your soil's ecosystem, dig a small hole and observe the number and variety of organisms present. If you find few or no signs of life, consider incorporating organic matter and reducing the use of chemical fertilizers and pesticides to encourage a healthy soil ecosystem.

You'll create a solid foundation for your no-till garden by assessing and improving your soil. With healthy soil in place, you'll be well on your way to enjoying the numerous benefits of no-till farming at home.

Choosing the Right Plants and Crop Rotation for Your No-Till Garden

Embarking on the no-till gardening journey requires a thoughtful approach to selecting the right plants and establishing a crop rotation system that promotes soil health and productivity. In this section, we will discuss the importance of plant selection and crop rotation and guide how to make the best choices for your no-till home garden.

When it comes to no-till gardening, not all plants are created equal. Some plants are better suited for this method due to their root systems, growth habits, and ability to improve soil structure. As you plan your no-till garden, consider the following factors:

- **Root Systems:** Plants with deep, fibrous root systems, such as legumes and grasses, are ideal for no-till gardens because they help to break up compacted soil and improve aeration. These plants also contribute to soil fertility by fixing nitrogen from the atmosphere.
- **Growth Habits:** Choose plants that grow well together and complement each other's needs. For example, tall plants like corn can provide shade for low-growing, shade-tolerant plants like lettuce. This not only maximizes space but also helps to maintain soil moisture and suppress weeds.
- **Soil Improvement:** Some plants, like legumes and certain cover crops, can improve soil fertility and structure. Including these plants in your no-till garden can help to create a healthier, more productive growing environment.

The Role of Crop Rotation in No-Till Gardening

Crop rotation is a crucial component of no-till gardening. It helps maintain soil health, prevent the buildup of pests and diseases, and ensure a balanced nutrient supply for your plants. Here are some fundamental principles to keep in mind when planning your crop rotation:

- **Grouping Plants by Family:** Organize your garden into sections based on plant families. This will make it easier to rotate crops each season, as plants within the same family often have similar nutrient requirements and are susceptible to the same pests and diseases.
- **Rotating Nutrient Demands:** Different plants have varying nutrient needs, and continuously planting the same crop in the same location can deplete the soil of essential nutrients. By rotating crops with different nutrient demands, you can help to maintain a balanced nutrient supply in your soil.
- **Breaking Pest and Disease Cycles:** Pests and diseases often target specific plant families. By rotating crops, you can disrupt the life cycles of these pests and diseases, reducing their impact on your garden.

Creating a Crop Rotation Plan

To create an effective crop rotation plan for your no-till garden, follow these steps:

1. Divide your garden into sections based on plant families.
2. Determine the nutrient needs of each plant family and group them accordingly (e.g., heavy feeders, light feeders, and soil builders).
3. Create a rotation schedule that allows each section to be planted with a different plant family each season, ensuring that light feeders or soil builders follow heavy feeders.
4. Keep a record of your crop rotation plan to help you track your garden's progress and make adjustments as needed.

By carefully selecting the right plants and implementing a thoughtful crop rotation plan, you can create a thriving, sustainable, no-till garden that benefits both your plants and the environment. In the next section, we will explore the use of organic mulching and cover crops to further enhance soil health in your no-till home garden.

Implementing Organic Mulching and Cover Crops for Soil Health

One of the essential aspects of a successful no-till home garden is maintaining the health and fertility of your soil. This section will delve into the importance of organic mulching and cover crops, two key components that contribute to the overall well-being of your garden's foundation.

Organic Mulching: A Natural Blanket for Your Soil

Organic mulching is a simple yet highly effective technique that involves spreading a layer of organic material, such as leaves, straw, or wood chips, over the surface of your garden soil. This *"natural blanket"* serves multiple purposes, all of which contribute to your no-till garden's overall health and productivity.

- **Soil Temperature Regulation:** Mulch acts as an insulator, helping to keep the soil cool during hot summer days and warm during cooler nights. This temperature regulation is crucial for developing healthy root systems and overall plant growth.
- **Moisture Retention:** Mulch helps retain soil moisture by reducing evaporation, ensuring your plants have a consistent water supply. This is especially important in no-till gardens, as the soil structure remains undisturbed, allowing for better water infiltration and retention.
- **Weed Suppression:** A layer of organic mulch can effectively suppress the growth of weeds by blocking sunlight and making it difficult for weed seeds to germinate. This not only saves you time and effort in weeding but also eliminates the need for harmful chemical herbicides.
- **Nutrient Recycling:** As the organic mulch decomposes, it releases essential nutrients back into the soil, providing a slow and steady source of nourishment for your plants. This

natural fertilization process is vital for maintaining the fertility of your no-till garden.

Cover Crops: Nature's Soil Builders

Cover crops, also known as green manures, are plants that are grown specifically to improve soil health and fertility. These plants are typically sown between growing seasons or alongside your main crops and offer numerous benefits for your no-till garden.

- **Soil Erosion Prevention:** Cover crops help to protect your soil from erosion by anchoring it with their roots and shielding it from the impact of heavy rains.
- **Organic Matter Addition:** When cover crops are terminated and left on the soil surface, they decompose and contribute valuable organic matter to the soil. This organic matter provides nutrients and improves soil structure and water-holding capacity.
- **Nitrogen Fixation:** Some cover crops, such as legumes, can "fix" nitrogen from the atmosphere and convert it into a form that plants can use. This natural nitrogen source can significantly reduce the need for synthetic fertilizers in your no-till garden.
- **Pest Management:** Certain cover crops can help to suppress pests and diseases by attracting beneficial insects or acting as a natural deterrent to harmful organisms.

By incorporating organic mulching and cover crops into your no-till home garden, you are taking crucial steps toward creating a thriving, sustainable, and productive growing environment. In addition, these techniques contribute to your soil's overall health and promote a more eco-friendly and self-sufficient gardening approach.

Managing Pests and Weeds in a No-Till Home Garden

One of the most rewarding aspects of no-till gardening is the harmonious relationship it fosters between your plants and the natural environment. However, this balance can sometimes be disrupted by unwanted pests and weeds. In this section, we will explore effective strategies for managing these common garden challenges without compromising the integrity of your no-till garden.

Encouraging Beneficial Insects and Wildlife

A healthy no-till garden is teeming with life; many of these creatures can help keep pests in check. By planting a diverse range of flowers and herbs, you can attract beneficial insects like ladybugs, lacewings, and ground beetles, which feed on common garden pests. Similarly, birds and small mammals can also be valuable allies in your battle against unwanted insects. Encourage their presence by providing nesting boxes, bird baths, and natural habitats like brush piles or rock walls.

Physical and Mechanical Controls

Sometimes, a hands-on approach is necessary to manage pests and weeds in your no-till garden. Regularly inspect your plants for signs of infestation, and remove any affected leaves or stems by hand. For larger pests like slugs and snails, you can use barriers such as copper tape or diatomaceous earth to deter them from feasting on your plants. When it comes to weeds, a thick layer of organic mulch can help suppress their growth by blocking sunlight. If weeds do manage to break through, remove them by hand or with a sharp hoe, taking care not to disturb the soil structure.

Biological Controls

In some cases, introducing natural predators or pathogens can effectively manage pests in your no-till garden. For example, you can purchase and release beneficial nematodes to combat grubs or introduce Bacillus thuringiensis (Bt) to control caterpillars. Always research and choose biological controls that are specific to the pests you are targeting, and ensure they are safe for your garden's ecosystem.

Organic Pesticides and Herbicides

While the goal of no-till gardening is to minimize the use of chemicals, there may be instances where organic pesticides or herbicides are necessary. When selecting these products, look for options that are derived from natural sources and have minimal impact on the environment. Neem oil, for example, is an effective organic pesticide that can help control a variety of pests, while vinegar can be used as a non-toxic herbicide for the spot treatment of weeds. Always follow the manufacturer's instructions and apply these products judiciously to avoid harming beneficial insects and wildlife.

Preventative Measures

Prevention is always better than cure, and this holds true for managing pests and weeds in your no-till garden. By maintaining healthy soil, practicing proper crop rotation, and using organic mulching and cover crops, you can create an environment less conducive to pest infestations and weed growth. Additionally, selecting disease-resistant plant varieties and ensuring adequate spacing between plants can further reduce the likelihood of problems arising.

In conclusion, managing pests and weeds in a no-till home garden requires a combination of proactive and reactive strategies. By working with nature and employing various organic and mechanical controls, you can maintain a healthy, productive garden that aligns with the principles of no-till farming.

Reaping the Benefits

As we reach the end of our journey into the world of no-till home gardening, it's time to reflect on the numerous benefits that await you when you embrace this sustainable and rewarding approach to cultivating your food. By adopting the no-till philosophy, assessing your soil, choosing the right plants and crop rotation, implementing organic mulching and cover crops, and managing pests and weeds, you have laid the groundwork for a thriving, eco-friendly garden that will nourish both your body and soul.

First and foremost, no-till gardening promotes soil health, which is the cornerstone of a successful garden. By eliminating the need for tilling, you preserve the soil structure, reduce erosion, and allow beneficial organisms like earthworms and microbes to thrive. This, in turn, leads to improved nutrient availability and water retention, ensuring that your plants receive the sustenance they need to grow strong and healthy.

Moreover, no-till gardening helps to reduce your carbon footprint and combat climate change. Traditional tilling methods release significant amounts of carbon dioxide into the atmosphere, contributing to global warming. By adopting a no-till approach, you actively participate in the fight against climate change and do your part to protect our precious planet.

Another advantage of no-till home gardening is the reduced need for chemical fertilizers and pesticides. By nurturing a healthy soil ecosystem and practicing crop rotation, you create a natural defense against pests and diseases. This saves you money on costly chemicals and ensures that the food you grow is free from harmful residues, providing you and your family with the healthiest and most nutritious produce possible.

Furthermore, no-till gardening can save you time and energy. Traditional tilling methods can be labor-intensive and time-consuming. In contrast, no-till gardening allows you to focus on other aspects of garden maintenance and enjoyment. With less time spent on tilling, you can devote more attention to planting, harvesting, and savoring the fruits of your labor.

Lastly, no-till gardening fosters a connection with the earth and the natural processes that sustain life. By working in harmony with nature, you are cultivating a deeper appreciation for the delicate balance of our ecosystem and the importance of preserving it for future generations. This sense of stewardship and responsibility can be a powerful motivator and a source of immense satisfaction as you watch your garden flourish.

In conclusion, preparing your home garden for no-till farming is a rewarding and transformative experience that offers numerous benefits for you and the environment. By embracing this sustainable approach, you create a thriving garden and contribute to a healthier planet and a brighter future for all. So, go forth and reap the **bountiful rewards** of no-till home gardening, knowing that you are positively impacting the world around you.

Chapter Summary

1. No-till gardening is a sustainable and eco-friendly approach that prioritizes soil health and the environment, focusing on preserving and enhancing the soil's natural structure and vitality.
2. Assessing your soil type, pH level, drainage, fertility, and soil life is crucial for creating a successful no-till garden, as healthy soil directly impacts the success of your endeavor.
3. Choosing the right plants and establishing a crop rotation system promotes soil health and productivity while also preventing the buildup of pests and diseases.
4. Organic mulching and cover crops are essential components of no-till gardening, as they help maintain soil health, improve fertility, and prevent erosion.
5. Encouraging beneficial insects and wildlife, using physical and mechanical controls, biological controls, organic

pesticides and herbicides, and practicing preventative measures can help manage pests and weeds in a no-till garden.

6. No-till gardening promotes soil health, reduces carbon footprint, and decreases the need for chemical fertilizers and pesticides, resulting in healthier and more nutritious produce.
7. Adopting a no-till approach saves time and energy, allowing gardeners to focus on other aspects of garden maintenance and enjoyment.
8. No-till gardening fosters a sense of connection with the earth and the natural processes that sustain life, cultivating a deeper appreciation for the delicate balance of our ecosystem and the importance of preserving it for future generations.

5

SELECTING THE RIGHT CROPS FOR YOUR NO-TILL GARDEN

A farmer's market overflowing with fresh, colorful produce, from bright orange pumpkins to deep purple eggplants.

No-till gardening is an innovative and sustainable approach to cultivating your home garden. You can maintain a healthy ecosystem that fosters plant growth and productivity by eliminating the need for tilling and disturbing the soil. However, the success of your no-till garden largely depends on selecting the right crops that are well-suited to your specific garden conditions. In this chapter, we will explore the importance of choosing suitable crops for your no-till garden and provide you with the knowledge and tools to make informed decisions that will ultimately lead to a thriving and bountiful harvest.

Choosing the suitable crops for your no-till garden is not just about picking your favorite vegetables or flowers; it's about understanding each plant's unique characteristics and how they interact with the soil, climate, and other plants in your garden. By selecting crops that are well-adapted to your garden's conditions, you can ensure that your plants will grow strong and healthy, minimizing the need for additional inputs such as fertilizers and pesticides. Moreover, choosing suitable crops will help you maximize the benefits of no-till gardening, including improved soil structure, increased water retention, and enhanced nutrient cycling.

In addition to considering each crop's specific needs and preferences, it's essential to consider the overall composition of your no-till garden. Plant diversity plays a crucial role in creating a resilient and productive garden ecosystem. By incorporating a variety of crops with different growth habits, root structures, and nutrient requirements, you can promote a more balanced and efficient use of resources and encourage beneficial interactions between plants and soil organisms.

In the following sections, we will guide you through the process of assessing your garden's soil, climate, and space and provide you with top crop recommendations for a successful no-till garden. We will also discuss the importance of experimenting and adapting your crop selection as you learn more about your garden's unique conditions and challenges. By the end of this chapter, you will be well-equipped to cultivate a thriving no-till garden through brilliant crop selection, setting the stage for a bountiful and sustainable harvest season after season.

Assessing Your Garden's Soil, Climate, and Space

Before diving into the exciting world of no-till gardening, it's crucial to understand the unique characteristics of your garden. By assessing your garden's soil, climate, and space, you'll be better equipped to select the suitable crops that will thrive in your no-till garden. In this section, we'll guide you through the process of evaluating these factors and how they impact your crop selection.

First, let's talk about soil. The foundation of any successful garden lies in the quality of its soil. No-till gardening is particularly reliant on healthy soil, as it promotes the natural processes that enrich the soil without disturbing its structure. To assess your soil, start by observing its texture, color, and drainage capabilities. You can also conduct a simple soil test to determine its pH level and nutrient content. This information will help you choose crops that are well-suited for your soil type and guide you in making any necessary amendments.

Next, consider your **garden's climate.** Temperature, rainfall, and sunlight are all critical factors that influence the growth and success of your crops. Research the average temperatures and precipitation levels in your area, and be mindful of the length of your growing season. This will help you select crops that thrive in your specific climate conditions. Additionally, pay attention to the microclimate within your garden, as factors such as shade, wind, and proximity to structures can create unique growing conditions.

Lastly, evaluate the **space** available in your garden. No-till gardening often involves planting a diverse range of crops in close proximity to one another, which can lead to more efficient use of space. However, it's essential to ensure that each plant has enough room to grow and access sunlight. Take note of the dimensions of your garden and any existing structures or features that may impact the available space. This will help you determine the number and variety of crops you can plant and their ideal spacing.

In summary, understanding your garden's soil, climate, and space is the first step towards selecting the right crops for your no-till garden. By taking the time to assess these factors, you'll be well on your way to creating a thriving, sustainable garden that is uniquely tailored to your specific conditions. In the next section, we'll explore the benefits of plant diversity in a no-till garden and provide some top crop recommendations to help you get started.

The Benefits of Plant Diversity in a No-Till Garden

Plant diversity is a crucial aspect of any successful no-till garden. By cultivating various crops, you can make your garden more visually appealing and reap numerous benefits that contribute to a healthier, more productive, and more sustainable growing environment. This section will explore the many advantages of incorporating plant diversity in your no-till garden.

First and foremost, plant diversity helps to **improve soil health.** Different plants have unique root structures and nutrient requirements, which means that they contribute to and draw from the soil in different ways. Therefore, growing a diverse range of crops ensures that your soil remains rich in essential nutrients and maintains a balanced ecosystem. This, in turn, promotes the growth of beneficial microorganisms and reduces the need for synthetic fertilizers.

Another significant benefit of plant diversity is its role in pest management. Monoculture gardens, where only one crop type is grown, can become a breeding ground for pests and diseases. In contrast, a diverse garden makes it more difficult for pests to thrive, as they are less likely to find a continuous food source. Moreover, many plants have natural pest-repelling properties, which can help protect your crops without the need for harmful chemical pesticides.

Plant diversity also aids in pollination, which is vital for the successful growth and reproduction of many crops. Growing various flowering plants can attract pollinators, such as bees, butterflies, and birds, to your garden. These pollinators will not only help to increase the yield of your crops but also contribute to the ecosystem's overall health.

Furthermore, plant diversity can help to conserve water in your no-till garden. Different plants have varying water requirements and rates of water uptake. Growing a mix of crops with diverse water needs can optimize your garden's water usage and reduce the risk of overwatering or underwatering your plants. This is particularly important in regions where water scarcity is a concern.

Lastly, plant diversity allows you to experiment with various crops and discover which best suits your garden's unique conditions. You can learn from your successes and failures by trying different plant combinations and ultimately develop a thriving no-till garden tailored to your specific needs and preferences.

In conclusion, incorporating plant diversity in your no-till garden is aesthetically pleasing and essential for creating a healthy, productive, and sustainable growing environment. In addition, by carefully selecting a diverse range of crops, you can improve soil health, manage pests, conserve water, and optimize your garden's overall performance. So, as you plan your no-till garden, remember to embrace the power of plant diversity and enjoy its many benefits.

Top Crop Recommendations for a Successful No-Till Garden

This section will explore some of the best crops to grow in your no-till garden. These recommendations are based on their compatibility with no-till practices, ability to improve soil health, and overall ease of cultivation. However, remember that these crops' success will still depend on your garden's specific conditions, so be sure to consider your soil, climate, and available space when making your selections.

Legumes: Legumes, such as beans, peas, and lentils, are an excellent choice for no-till gardens due to their ability to fix nitrogen in the soil. This process enriches the soil and reduces the need for synthetic fertilizers, making legumes valuable to any no-till garden. Additionally, they are relatively easy to grow and can be planted in succession for a continuous harvest throughout the growing season.

Root Vegetables: Root vegetables like carrots, beets, and radishes are well-suited for no-till gardens because they can help break up compacted soil with their strong, penetrating roots. This natural soil aeration promotes healthy root growth for other plants and improves overall soil structure. Root vegetables are also versatile in the kitchen and can be stored for extended periods, making them a practical choice for home gardeners.

Leafy Greens: Leafy greens like lettuce, spinach, and kale are another excellent choice for no-till gardens. They have shallow root systems, so they won't disturb the soil structure as they grow. Additionally, they are typically fast-growing and can be harvested multiple times throughout the season, providing a consistent supply of fresh, nutritious greens for your table.

Perennial Crops: Perennial crops, like asparagus, rhubarb, and artichokes, are a fantastic addition to no-till gardens because they return year after year without the need for replanting. This saves you time and effort and helps maintain soil structure by minimizing disturbances. Perennial crops may take a few years to establish, but their long-term benefits make them worth the investment.

Cover Crops: While not typically considered a traditional garden crop, cover crops play a crucial role in no-till gardens. Plants like clover, vetch, and rye can be grown between your main crops to protect the soil, suppress weeds, and add valuable organic matter when cut down and left to decompose. This practice helps maintain soil fertility and structure, making cover crops essential to a successful no-till garden.

In conclusion, selecting the right crops for your no-till garden is crucial for its success. You can create a thriving, productive garden that benefits you and the environment by choosing a diverse mix of legumes, root vegetables, leafy greens, perennials, and cover crops. Remember to experiment and adapt your crop choices based on your garden's unique conditions. Then, you'll be well on your way to a flourishing no-till garden.

Experimenting and Adapting: Finding the Perfect Crop Combination

As you embark on your no-till gardening journey, it's essential to remember that every garden is unique, and what works for one may not necessarily work for another. Therefore, it's crucial to experiment with different crops and adapt your planting strategies based on your garden's specific needs. This section will explore the importance of experimentation and adaptation in finding the perfect crop combination for your no-till garden.

One of the most exciting aspects of no-till gardening is the opportunity to discover new plant varieties and combinations that thrive in your specific environment. While starting with tried-and-true crop recommendations is helpful, it's equally important to be open to experimentation. By doing so, you'll learn more about your garden's unique characteristics and develop a deeper understanding of the plants that flourish within it.

When experimenting with new crops, it's essential to keep detailed records of your observations. Take note of factors such as germination rates, growth patterns, pest and disease resistance, and overall plant health. This information will prove invaluable as you continue to refine your no-till garden and adapt your planting strategies to suit your garden's needs better.

As you gain experience and knowledge, don't be afraid to **adjust your crop selection.** If a particular plant variety isn't performing well in your garden, consider replacing it with another plant that may be better suited to your soil, climate, and space. Additionally, pay attention to how different crops interact with one another. Some plants may complement each other, promoting healthier growth and improved pest resistance. In contrast, others may compete for resources, leading to stunted growth or other issues.

Incorporating cover crops and crop rotation into your no-till garden can also contribute to a more successful and diverse planting strategy. Cover crops like clover or rye can help improve soil health, suppress weeds, and provide valuable nutrients to your primary crops. Crop rotation, or changing the location of specific crops each year, can help prevent the buildup of pests and diseases while promoting better nutrient management.

In conclusion, finding the perfect crop combination for your no-till garden is an ongoing process that requires patience, observation, and a willingness to adapt. By experimenting with different crops, paying attention to their performance, and making adjustments as needed, you'll be well on your way to cultivating a thriving, diverse, and productive no-till garden.

Cultivating a Thriving No-Till Garden Through Smart Crop Selection

In conclusion, the key to a flourishing no-till garden lies in carefully selecting crops well-suited to your garden's unique conditions. By taking the time to understand your garden's soil, climate, and space, you can make informed decisions about which crops will thrive and contribute to a healthy, productive no-till garden.

Embracing plant diversity is another essential aspect of successful no-till gardening. By incorporating a wide variety of crops, you can create a more resilient garden ecosystem less susceptible to pests, diseases, and other challenges. Moreover, diverse plantings can lead to a more bountiful harvest, offering you a wide array of delicious and nutritious produce to enjoy.

Our top crop recommendations are a helpful starting point for your no-till garden, but don't be afraid to experiment and adapt your crop selection based on your personal preferences and experiences. Gardening is a continuous learning process, and by paying close attention to your garden's performance, you can fine-tune your crop selection over time to create the perfect combination for your specific needs.

In summary, smart crop selection is the cornerstone of a thriving no-till garden. You can cultivate a bountiful and sustainable garden that benefits you and the environment by choosing crops that are well-adapted to your garden's conditions, embracing plant diversity, and continually experimenting and adapting. So, go ahead and embark on your no-till gardening journey, and watch as your garden flourishes with life and abundance.

Chapter Summary

1. No-till gardening is a sustainable approach that relies on selecting the right crops suited to your garden's specific conditions for success.
2. Assessing your garden's soil, climate, and space is crucial for choosing suitable crops that thrive in a no-till garden.
3. Plant diversity is vital in creating a resilient and productive garden ecosystem, improving soil health, pest management, pollination, and water conservation.
4. Top crop recommendations for no-till gardens include legumes, root vegetables, leafy greens, perennial crops, and cover crops.
5. Experimentation and adaptation are essential in finding the perfect crop combination for your no-till garden, as every garden is unique.
6. Keeping detailed records of your observations and adjusting your crop selection based on performance can help refine your no-till garden over time.
7. Incorporating cover crops and crop rotation can contribute to a more successful and diverse planting strategy in a no-till garden.
8. Cultivating a thriving no-till garden requires smart crop selection, embracing plant diversity, and continuous learning through experimentation and adaptation.

6

IMPLEMENTING COVER CROPS AND CROP ROTATION STRATEGIES

A field of lush green cover crops growing tall under the sun.

No-till farming has gained significant traction among home gardeners and small-scale farmers, and for a good reason. This sustainable approach to agriculture conserves soil and water resources and promotes a healthy and thriving ecosystem within your garden. One of the key components of successful no-till farming is the strategic use of cover crops and crop rotation. In this chapter, we will delve into the importance of these practices and provide practical guidance on implementing them in your home garden.

Cover crops and crop rotation go hand-in-hand in creating a robust and resilient no-till farming system. Cover crops are plants grown primarily for the benefit of the soil rather than for harvest. They play a crucial role in protecting the soil surface from erosion, suppressing weeds, and enhancing soil fertility. Covering the soil with a living "blanket" of plants essentially mimics the natural processes that occur in undisturbed ecosystems, where the soil is rarely left bare and exposed to the elements.

Crop rotation, on the other hand, is the practice of growing different types of crops in the same area over a series of growing seasons. This time-tested technique helps break pest and disease cycles and improves soil structure and nutrient availability. By rotating crops, you are essentially "confusing" pests and pathogens, making it more difficult for them to establish a foothold in your garden. Moreover, different crops have varying nutrient requirements and root structures, which can help to balance the nutrient profile of your soil and promote a diverse community of soil organisms.

Together, cover crops and crop rotation form the **backbone** of a successful no-till farming system. By implementing these practices, you are taking essential steps towards creating a more sustainable and productive garden that will reward you with bountiful harvests for years to come. In the following sections, we will explore choosing the right cover crops for your home garden, effective crop rotation techniques for soil health and pest management, and integrating these strategies into your no-till farming plan. So, let's embark on this exciting journey toward a more sustainable and fruitful home garden!

Choosing the Right Cover Crops for Your Home Garden

Selecting the perfect cover crops for your home garden is essential in implementing no-till farming practices. Cover crops not only protect and enrich your soil, but they also play a crucial role in maintaining a healthy garden ecosystem. In this section, we will explore the various factors to consider when choosing cover crops and offer some popular options to help you make the best decision for your garden.

When selecting cover crops, it's important to consider your garden's specific needs and goals. Here are a few factors to keep in mind:

Climate and Growing Season: Choose cover crops that are well-suited to your local climate and can thrive within your garden's growing season. Some cover crops, such as winter rye and hairy vetch, are cold-hardy and can be grown during the cooler months. Others, like buckwheat and cowpeas, prefer warmer temperatures.

Soil Type and Nutrient Needs: Different cover crops have varying preferences for soil type and nutrient levels. For instance, legumes like clover and peas can fix nitrogen from the atmosphere, making them ideal for nitrogen-deficient soils. On the other hand, deep-rooted cover crops like daikon radish can help break up compacted soil and improve drainage.

Pest and Disease Management: Some cover crops can help suppress weeds, attract beneficial insects, and deter pests. For example, marigolds release a chemical that repels nematodes, while buckwheat attracts pollinators and predatory insects that help control pests.

Garden Layout and Space: Consider the size and layout of your garden when selecting cover crops. If you have limited space, opt for low-growing cover crops like white clover or creeping thyme that can be easily incorporated into your existing garden beds. For larger gardens, you can use taller cover crops like sunflowers or sorghum to create natural barriers or windbreaks.

Now that you know what factors to consider, let's explore some popular cover crop options for your home garden:

Legumes: These nitrogen-fixing plants, such as clover, peas, and beans, are excellent choices for improving soil fertility. They can be easily incorporated into your crop rotation plan and provide additional benefits like attracting pollinators and providing a habitat for beneficial insects.

Grasses and Grains: Crops like winter rye, barley, and oats are ideal for preventing soil erosion, suppressing weeds, and adding organic matter to the soil. They can also serve as a natural mulch when left on the soil surface after they've been mowed or crimped.

Brassicas and Mustards: These cover crops, including radishes, turnips, and mustard greens, are known for suppressing soil-borne pests and diseases. They also help improve soil structure and fertility by breaking up compacted soil with their deep roots.

Flowering Cover Crops: Plants like buckwheat, phacelia, and sunflowers not only add beauty to your garden but also attract pollinators and beneficial insects. They can be used as a living mulch or incorporated into your garden beds to improve soil fertility and structure.

By carefully selecting the right cover crops for your home garden, you'll be well on your way to reaping the numerous benefits of no-till farming. The following section will explore effective crop rotation techniques to further enhance your garden's health and productivity.

Effective Crop Rotation Techniques for Soil Health and Pest Management

In no-till farming, crop rotation is crucial in maintaining soil health and managing pests. By alternating the types of crops grown in a specific area, you can prevent the depletion of essential nutrients and disrupt the life cycles of pests and diseases. In this section, we will explore effective crop rotation techniques that will help you create a thriving and sustainable home garden.

To begin with, it is essential to understand the different crop families and their specific needs. Crops are generally classified into four main groups: legumes, brassicas, root crops, and leafy greens. Each group has unique nutrient requirements and pest vulnerabilities. Systematically rotating these groups ensures that your soil remains fertile and your plants stay healthy.

Here are some steps to follow for effective crop rotation:

1. **Divide your garden into sections:** Start by dividing your garden into four equal sections, each dedicated to one of the four crop groups. This will make it easier to plan and manage your crop rotation.
2. **Rotate clockwise or counterclockwise:** Each year, move each crop group to the next section in either a clockwise or counterclockwise direction. This ensures that each group is grown in a different section for at least three years before returning to its original location. This time gap helps to break the life cycles of pests and diseases and allows the soil to recover from any nutrient depletion.
3. **Plant cover crops between rotations:** To further enhance soil health and pest management, consider planting cover crops between rotations. Cover crops, such as clover, rye, or buckwheat, can help to suppress weeds, improve soil structure, and provide additional nutrients for your main crops.
4. **Keep records:** Maintaining accurate records of your crop rotation plan is essential for long-term success. Note down the crops grown in each section, the dates of planting and harvesting, and any pest or disease issues that arise. This information will help refine your rotation plan and make informed decisions about future plantings.
5. **Be flexible:** While following a systematic crop rotation plan is important, it is equally important to be flexible and adapt to changing conditions. If you notice a particular pest or disease becoming a problem, consider adjusting your rotation schedule or introducing additional crop groups to help combat the issue.

In conclusion, effective crop rotation is vital to no-till farming at home. By understanding the needs of different crop groups and implementing a systematic rotation plan, you can create a healthy and productive garden that is resistant to pests and diseases. Remember to monitor your garden closely, keep accurate records, and be prepared to adapt your strategies as needed for long-term success.

Integrating Cover Crops and Crop Rotation into Your No-Till Farming Plan

Incorporating cover crops and crop rotation into your no-till farming plan is crucial for maintaining soil health, controlling pests, and ensuring a bountiful harvest. By understanding the symbiotic relationship between these two practices, you can create a thriving ecosystem in your home garden. This section will explore how to effectively integrate cover crops and crop rotation into your no-till farming plan.

First, let's take a closer look at the role of cover crops in no-till farming. Cover crops are plants grown primarily for the benefit of the soil rather than for harvest. They protect the soil from erosion, suppress weeds, and improve soil structure by adding organic matter. Some cover crops, like legumes, can also fix nitrogen in the soil, reducing the need for synthetic fertilizers.

To integrate cover crops into your no-till farming plan, start by selecting the right cover crop for your garden's specific needs. Consider factors such as climate, soil type, and the primary crops you plan to grow. For instance, planting legumes like clover or peas can help **replenish nitrogen levels** if you have nitrogen-deficient soil. Once you've chosen the appropriate cover crop, plant it either between your main crops or during the off-season.

Now, let's turn our attention to crop rotation. Crop rotation is the practice of growing different types of crops in the same area in sequential seasons. This method helps prevent soil depletion, as different plants require varying nutrients and can help balance the soil's nutrient levels. Additionally, crop rotation can disrupt the life cycles of pests and diseases, reducing their impact on your garden.

To integrate crop rotation into your no-till farming plan, divide your garden into sections or beds. Allocate each section to a specific crop family, such as legumes, brassicas, or root vegetables. Then, rotate these sections each growing season, ensuring that the same crop family is not planted in the same area for at least three years. This will help maintain soil fertility and reduce the risk of pests and diseases.

Combining cover crops and crop rotation in your no-till farming plan can lead to a more resilient and productive garden. For example, you might plant a nitrogen-fixing cover crop like clover in a section that previously hosted heavy-feeding crops like tomatoes or corn. This will replenish the soil's nitrogen levels, suppress weeds, and improve soil structure for the next crop in the rotation.

In conclusion, integrating cover crops and crop rotation into your no-till farming plan is essential for creating a sustainable and thriving home garden. By carefully selecting cover crops and implementing a well-planned crop rotation strategy, you can maintain soil health, manage pests, and ensure a bountiful harvest for years to come. Remember to monitor and adapt your strategies as needed, and enjoy the lasting benefits of these eco-friendly gardening practices.

Monitoring and Adapting Your Strategies for Long-Term Success

In the world of no-till farming, it's essential to remember that your garden is a living, ever-changing ecosystem. As you implement cover crops and crop rotation strategies, monitoring your garden's progress and adapting your techniques to ensure long-term success is crucial. In this section, we'll discuss how to monitor your garden's health, make necessary adjustments, and ultimately create a thriving no-till home garden.

First and foremost, it's important to regularly observe your garden. Pay close attention to the growth and health of your plants, the presence of pests, and the overall condition of your soil. By staying in tune with your garden, you'll be better equipped to identify any issues that may arise and address them promptly.

One key aspect of monitoring your garden's health is conducting **periodic soil tests.** These tests will provide valuable information about your soil's nutrient levels, pH balance, and organic matter content. By analyzing the results, you can make informed decisions about which cover crops to plant and how to adjust your crop rotation plan to maintain optimal soil health.

Another essential component of monitoring your no-till garden is keeping a garden journal. Documenting your observations, successes, and challenges will help you track your progress and make data-driven decisions. Be sure to include information about the cover crops you've planted, the crop rotation schedule you've followed, and any pest or disease issues you've encountered. Over time, you'll be able to identify patterns and trends that will inform your future decisions.

As you gather information about your garden, don't be afraid to make adjustments to your cover crop and crop rotation strategies. No-till farming is an ongoing learning process, and being flexible and open to change is essential. For example, consider trying a different species or variety if a particular cover crop isn't performing well in your garden. Similarly, if you find that a specific crop rotation sequence isn't effectively managing pests or improving soil health, experiment with alternative rotations.

In addition to making adjustments based on your observations, staying informed about the latest research and best practices in no-till farming is important. Joining local gardening clubs, attending workshops, and connecting with fellow no-till enthusiasts can provide valuable insights and inspiration for refining your strategies.

In conclusion, monitoring and adapting your cover crop and crop rotation strategies are essential components of long-term success in no-till farming. By staying observant, conducting regular soil tests, keeping a garden journal, and being open to change, you'll be well on your way to creating a thriving, sustainable home garden that benefits you and the environment.

The Lasting Benefits of Implementing Cover Crops and Crop Rotation in Your Home Garden

As we reach the end of this enlightening journey through the world of no-till farming, it is essential to reflect on the lasting benefits that cover crops and crop rotation strategies can bring to your home garden.

When combined with the no-till approach, these practices can create a thriving, sustainable, and productive garden that will benefit you and your family and contribute to a healthier environment.

By incorporating cover crops into your garden, you are taking a proactive step towards improving soil health, reducing erosion, and enhancing plant nutrient availability. These hardworking plants act as living mulch, protecting the soil from the elements and providing a habitat for beneficial insects and microorganisms. In addition, as they decompose, cover crops release valuable nutrients into the soil, ensuring that your garden remains fertile and productive for years to come.

Crop rotation, on the other hand, is a time-tested technique that helps prevent the buildup of pests and diseases while promoting a balanced nutrient cycle in your garden. By rotating your crops, you are effectively breaking the life cycles of pests and diseases, reducing the need for chemical interventions, and promoting a more diverse and resilient ecosystem. This practice also ensures that your soil's nutrients are used efficiently, allowing you to grow a wide variety of crops without depleting the soil's resources.

When combined, cover crops and crop rotation strategies create a powerful synergy that enhances your no-till garden's overall health and productivity. By integrating these practices into your gardening routine, you are investing in your garden's long-term success and contributing to a more sustainable and environmentally friendly approach to food production.

As you continue to explore the world of no-till farming and refine your techniques, remember that the key to success lies in monitoring and adapting your strategies to suit your garden's unique needs. Be open to learning from your experiences, and don't be afraid to make adjustments as needed. With patience, dedication, and a willingness to embrace new ideas, you can create a thriving, sustainable, and productive home garden that will serve as a testament to the power of no-till farming and the lasting benefits of implementing cover crops and crop rotation strategies.

In conclusion, the journey towards a more sustainable and productive home garden begins with understanding and implementing the principles of no-till farming, cover crops, and crop rotation. By embracing these practices, you are taking a significant step towards creating a garden that **nourishes your family** and contributes to a **healthier planet.** So, go ahead and sow the seeds of change, and watch as your garden blossoms into a living example of the lasting benefits of no-till farming.

Chapter Summary

1. No-till farming is a sustainable approach to agriculture that conserves soil and water resources while promoting a healthy ecosystem. Cover crops and crop rotation are key components of successful no-till farming.
2. Cover crops protect the soil from erosion, suppress weeds, and enhance soil fertility. They can also help manage pests and diseases and improve soil structure.
3. Crop rotation involves growing different types of crops in the same area over a series of growing seasons. This technique helps break pest and disease cycles and improves soil structure and nutrient availability.
4. Choosing the right cover crops for your home garden involves considering factors such as climate, soil type, nutrient needs, pest management, and garden layout.
5. Effective crop rotation requires dividing your garden into sections, rotating crops clockwise or counterclockwise, planting cover crops between rotations, keeping records, and being flexible.
6. Integrating cover crops and crop rotation into your no-till farming plan can lead to a more resilient and productive garden. This involves selecting appropriate cover crops, implementing a well-planned crop rotation strategy, and monitoring and adapting your strategies as needed.

7. Monitoring and adapting your cover crop and crop rotation strategies are essential for long-term success in no-till farming. This includes observing your garden, conducting regular soil tests, keeping a garden journal, and staying informed about the latest research and best practices.
8. Implementing cover crops and crop rotation in your home garden provides lasting benefits, including improved soil health, reduced erosion, enhanced nutrient availability, and a more sustainable and environmentally friendly approach to food production.

7

MANAGING WEEDS AND PESTS IN A NO-TILL GARDEN

A lush garden bursting with vibrant vegetables of every color and shape, their leafy greens reaching towards the sky.

No-till gardening is a sustainable, eco-friendly approach to cultivating a thriving, productive garden. By eliminating the need for tilling and minimizing soil disturbance, this method promotes healthy soil structure, conserves water, and reduces erosion. However, as with any gardening technique, no-till gardens are not immune to the challenges of weed and pest invasions. Therefore, effective and sustainable weed and pest management practices are key to maintaining a flourishing no-till garden.

This chapter will explore various natural, chemical-free, and biological methods for managing weeds and pests in your no-till garden. These techniques protect your plants, preserve soil health, and promote a balanced ecosystem that supports beneficial organisms. By understanding the importance of sustainable weed and pest management, you will be well-equipped to maintain a thriving, productive, and environmentally friendly no-till garden.

Whether you are a seasoned gardener or a beginner, this chapter will provide you with valuable insights and practical tips for managing weeds and pests in your no-till garden. So, let's dive into the world of sustainable weed and pest management and discover how to keep your no-till garden healthy and thriving.

Natural Weed Control Techniques for No-Till Gardens

This section will explore various natural weed control techniques that effectively maintain a healthy and thriving no-till garden. These methods are not only environmentally friendly but also help in preserving the soil structure and promote biodiversity in your garden. In addition, implementing these techniques allows you to enjoy a beautiful and productive garden without resorting to harmful chemicals or disruptive tilling practices.

Mulching: One of the most effective and widely used natural weed control techniques is mulching. Covering the soil surface with a layer of organic materials such as straw, wood chips, or leaves can suppress weed growth by blocking sunlight and making it difficult for weeds to take root. Mulching also helps retain soil moisture, moderating soil temperature and adding nutrients to the soil as the organic matter decomposes over time.

Planting cover crops: Cover crops are plants that are grown primarily for their ability to suppress weeds, improve soil fertility, and prevent soil erosion. By planting cover crops like clover, vetch, or rye in your no-till garden, you can effectively outcompete weeds for sunlight, water, and nutrients. Additionally, cover crops provide a habitat for beneficial insects and microorganisms that contribute to a healthy garden ecosystem.

Solarization: Solarization is a natural weed control technique that involves using the sun's energy to heat the soil and kill weed seeds and pathogens. To solarize your garden, simply cover the soil with a clear plastic sheet during the hottest part of the summer. The trapped heat will effectively sterilize the top layer of soil, making it inhospitable for weed seeds to germinate.

Hand pulling and hoeing: While it may be labor-intensive, hand pulling and hoeing are tried and true methods of natural weed control. Regularly removing weeds from your no-till garden prevents them from competing with your desired plants for resources. Be sure to remove weeds before they have a chance to produce seeds, as this will help reduce the overall weed population in your garden.

Companion planting: Companion planting is the practice of growing certain plants together to support each other's growth and deter pests. By strategically placing plants with natural weed-suppressing properties near your desired crops, you can help keep weeds at bay. For example, planting aromatic herbs like basil, mint, or rosemary can help deter weeds and pests while providing a bountiful harvest.

Mowing and cutting: In larger no-till gardens or areas with perennial weeds, mowing or cutting the weeds before they have a chance to produce seeds can be an effective control method. This technique is particularly useful for managing grassy weeds and can help maintain a tidy and well-manicured garden.

Incorporating these natural weed control techniques into your no-till garden management plan allows you to enjoy a healthy, productive, and sustainable garden free from harmful chemicals and disruptive tilling practices. In the next section, we will delve into biological pest control methods that can further enhance the health and resilience of your no-till garden.

Biological Pest Control Methods for a Healthy No-Till Garden

In this section, we will explore the fascinating world of biological pest control methods, which harness the power of nature to keep your no-till garden healthy and thriving. By encouraging the presence of beneficial organisms and creating a balanced ecosystem, you can effectively manage pests without resorting to harmful chemicals. Let's dive into some of the most effective biological pest control methods for your no-till garden.

Beneficial Insects: Nature's Pest Control Squad

One of the most effective ways to control pests in your no-till garden is by attracting and nurturing beneficial insects. These tiny heroes, such as ladybugs, lacewings, and parasitic wasps, feed on common garden pests like aphids, caterpillars, and whiteflies. To attract these beneficial insects, plant a variety of flowering plants, especially those with small flowers like dill, fennel, and yarrow. You can also purchase and release these insects into your garden, but be sure to research the best time and method for doing so.

Birds: Feathered Friends and Foes of Pests

Birds can be a valuable ally in your quest for a healthy no-till garden, as they feed on a wide range of pests, including slugs, snails, and insects. To attract birds to your garden, provide them with nesting sites, such as birdhouses, and a source of freshwater. Planting a variety of trees and shrubs will also provide a natural habitat and food source for birds. However, be aware that some birds may also feast on your fruits and vegetables, so use bird netting to protect your crops when necessary.

Nematodes: Microscopic Warriors

Beneficial nematodes are microscopic roundworms that can help control pests in your no-till garden. These tiny organisms attack and kill pests like grubs, root maggots, and cutworms by entering their bodies and releasing bacteria that cause a fatal infection. You can purchase beneficial nematodes online or at garden centers and apply them to your garden according to the package instructions. Be sure to choose the right species of nematode for the pests you're targeting, as different species have different prey preferences.

Bats: Nighttime Pest Patrol

Bats may have a spooky reputation, but they're actually fantastic pest controllers in the garden. Many bat species feed on insects, including mosquitoes and moths, which can harm your plants. To attract bats to your garden, consider installing a bat house and planting night-blooming flowers to provide a food source. Remember that bats are a protected species in many areas, so be sure to follow local guidelines when installing a bat house.

Frogs and Toads: Slugs' Worst Nightmare

Frogs and toads are another excellent addition to your no-till garden's pest control team. These amphibians feed on a variety of pests, including slugs, snails, and insects. To create a welcoming habitat for frogs and toads, provide a shallow water source and plenty of hiding spots, such as rocks, logs, or toad houses.

By incorporating these biological pest control methods into your no-till garden, you can create a thriving, balanced ecosystem that keeps pests in check naturally. Not only will this help protect your plants and the environment, but it will also provide you with a fascinating glimpse into the complex web of life that exists in your garden.

Chemical-Free Pest Management Strategies

In this section, we will explore various chemical-free pest management strategies that are not only effective but also environmentally friendly. These methods are essential for maintaining the integrity of your no-till garden, as they help to preserve the soil structure and promote a healthy ecosystem. By adopting these sustainable practices, you can ensure that your garden remains productive and free from harmful pests without resorting to synthetic chemicals.

Physical Barriers: One of the simplest ways to deter pests from your no-till garden is by using physical barriers. This can include installing floating row covers, which are lightweight fabrics that protect your plants from insects while still allowing sunlight and air to pass through. Another option is to use copper tape or mesh around the base of your plants to prevent slugs and snails from reaching them. These barriers can be highly effective in keeping pests at bay without using chemicals.

Companion Planting: This strategy involves planting certain plants together to help deter pests or attract beneficial insects that prey on harmful pests. For example, planting marigolds near your tomato plants can help repel nematodes, while planting basil alongside your peppers can deter aphids. By carefully selecting and positioning your plants, you can create a natural defense system that protects your no-till garden from pests.

Beneficial Insects: Encouraging the presence of beneficial insects in your no-till garden is another effective chemical-free pest management strategy. Ladybugs, lacewings, and parasitic wasps are just a few examples of insects that prey on common garden pests. You can attract these helpful insects by planting flowers such as yarrow, dill, and fennel, which provide them with nectar and pollen. Additionally, consider installing insect hotels or nesting boxes to provide shelter for these beneficial insects.

Traps and Lures: Using traps and lures is another chemical-free method to control pests in your no-till garden. Sticky traps, for instance, can be used to catch flying insects such as whiteflies and aphids. You can also create homemade traps using simple materials like cardboard and duct tape to capture pests like slugs and snails. Additionally, pheromone traps can be employed to lure and trap specific pests, such as the codling moth, which is a common pest of fruit trees.

Manual Removal: For smaller no-till gardens, manual removal of pests can be an effective and chemical-free solution. Regularly inspect your plants for signs of infestation, and remove any pests you find by hand. This method works best for larger pests like caterpillars and beetles. Be sure to wear gloves and use a small container to collect the pests before disposing of them away from your garden.

By incorporating these chemical-free pest management strategies into your no-till gardening routine, you can maintain a healthy and productive garden while minimizing your impact on the environment. In addition, these sustainable practices protect your plants and contribute to the overall well-being of your garden's ecosystem, ensuring that your no-till garden thrives for years to come.

Monitoring and Maintaining a Weed and Pest-Free No-Till Garden

A thriving no-till garden is a beautiful sight to behold, but it requires consistent monitoring and maintenance to ensure that weeds and pests don't take over. This section will explore some practical tips and techniques to help you keep your no-till garden healthy and productive without resorting to harmful chemicals or back-breaking labor.

First and foremost, it's essential to regularly inspect your garden for signs of weed and pest infestations. This may seem like a simple task, but it's one that many gardeners overlook. By taking the time to walk through your garden and observe the plants closely, you can identify potential problems early on and take appropriate action before they become unmanageable.

When monitoring your garden, **watch out** for any unusual changes in plant growth, color, or overall health. These can be indicators of weed or pest issues. Additionally, pay attention to the presence of both beneficial and harmful insects and signs of their activity, such as chewed leaves or damaged stems.

One effective way to manage weeds in a no-till garden is through the use of mulch. Mulching not only helps to conserve moisture and regulate soil temperature, but it also acts as a barrier to prevent weed seeds from germinating. Organic mulches, such as straw, wood chips, or grass clippings, can be especially beneficial, as they break down over time and add valuable nutrients to the soil.

Another essential aspect of maintaining a weed and pest-free no-till garden is to promote a healthy ecosystem that supports beneficial insects and other natural predators. You can create a balanced environment that keeps pests in check by planting a diverse range of crops and incorporating flowering plants that attract pollinators and predatory insects.

It's also crucial to practice good **garden hygiene** to prevent the spread of pests and diseases. This includes removing dead plant material, which can harbor pests and diseases, and keeping the garden free of debris that can provide hiding places for unwanted critters. Additionally, be sure to sanitize your gardening tools regularly, especially if you've been working with plants that show signs of disease or pest infestations.

In some cases, manual removal of weeds and pests may be necessary. This can be done by hand-pulling weeds or using a hoe to cut them off at the soil level. Likewise, you can remove pests by hand or use a gentle stream of water to dislodge them from your plants. Remember, the key is to address these issues early before they have a chance to multiply and cause significant damage.

In conclusion, monitoring and maintaining a weed and pest-free no-till garden is an ongoing process that requires vigilance and dedication. By staying proactive and employing sustainable, chemical-free strategies, you can enjoy the many benefits of no-till gardening while minimizing the impact on the environment and your health.

The Importance of Sustainable Weed and Pest Management in No-Till Gardening

In conclusion, sustainable weed and pest management are essential to maintaining a thriving no-till garden. As we have explored throughout this chapter, adopting natural, biological, and chemical-free strategies not only contributes to the overall health of your garden but also promotes a more sustainable and eco-friendly approach to gardening.

By implementing the various techniques discussed in this chapter, such as mulching, crop rotation, and introducing beneficial insects, you are taking a proactive approach to weed and pest control. These methods not only prevent the need for harmful chemicals but also work in harmony with nature, supporting the delicate balance of your garden's ecosystem.

One of the key takeaways from this chapter is the importance of regular monitoring and maintenance in your no-till garden. By staying vigilant and addressing potential issues early on, you can prevent weeds and pests from becoming overwhelming problems. This proactive approach ensures the long-term success of your no-till garden. It allows you to enjoy the many benefits it offers, such as improved soil health, increased biodiversity, and a more sustainable way of growing your own food.

As a no-till gardener, you are part of a growing movement that recognizes the importance of working with nature rather than against it. By embracing sustainable weed and pest management practices, you are not only contributing to the health of your garden but also setting an example for others to follow. In doing so, you are playing a vital role in promoting a more environmentally conscious approach to gardening, one that has the potential to make a lasting impact on our planet.

In summary, sustainable weed and pest management are crucial for the success of your no-till garden and the well-being of the environment. By adopting the techniques discussed in this chapter, you can enjoy a thriving, healthy garden while also contributing to a more sustainable and eco-friendly future for us all. So, as you venture into the world of no-till gardening, remember that your efforts in responsibly managing weeds and pests will benefit your garden and the world around you. Happy gardening!

Chapter Summary

1. No-till gardening is a sustainable and eco-friendly approach to cultivating a thriving, productive garden, but it requires effective weed and pest management practices to maintain its health and productivity.
2. Natural weed control techniques, such as mulching, planting cover crops, solarization, hand pulling, hoeing, companion planting, and mowing, can help maintain a healthy no-till garden without resorting to harmful chemicals or disruptive tilling practices.
3. Using biological pest control methods, and attracting beneficial insects, birds, nematodes, bats, and frogs, can help create a balanced ecosystem that keeps pests in check naturally, protecting plants and the environment.
4. Chemical-free pest management strategies, such as physical barriers, companion planting, beneficial insects, traps, lures, and manual removal, can maintain a healthy and productive garden while minimizing the impact on the environment.
5. Regular monitoring and maintenance are essential for keeping a no-till garden weed and pest-free, including inspecting plants for signs of infestation, promoting a healthy ecosystem, and practicing good garden hygiene.
6. Mulching is crucial to maintaining a weed-free, no-till garden, as it conserves moisture, regulates soil temperature, and acts as a barrier to prevent weed seeds from germinating.
7. Sustainable weed and pest management practices contribute to the overall health of a no-till garden and promote a more sustainable and eco-friendly approach to

gardening, working in harmony with nature and supporting the delicate balance of the garden's ecosystem.

8. No-till gardeners play a vital role in promoting a more environmentally conscious approach to gardening by embracing sustainable weed and pest management practices, setting an example for others to follow, and contributing to a more sustainable and eco-friendly future for all.

8

FERTILIZATION AND NUTRIENT MANAGEMENT IN NO-TILL FARMING

A lush garden bursting with vibrant vegetables of every color and shape, their leafy greens reaching towards the sky.

Welcome to Chapter 8, where we delve into the fascinating world of fertilization and nutrient management in no-till farming. As a gardener or small-scale farmer, you're likely aware that providing your plants with the right nutrients is essential for their growth and overall health. However, when it comes to no-till farming, the approach to fertilization and nutrient management is quite different from that of conventional farming methods.

In this chapter, we'll explore the unique aspects of maintaining soil fertility in a no-till garden and how you can ensure that your plants receive the nutrients they need to thrive. We'll begin by discussing the importance of soil health and understanding the nutrient needs of your plants. From there, we'll introduce you to a variety of organic fertilizers and amendments that are well-suited for no-till gardens, as well as efficient application techniques that minimize disturbance to the soil structure.

As you progress through this chapter, you'll learn how to monitor and adjust nutrient levels in your no-till garden to optimize plant growth and, ultimately, achieve sustainable fertility that benefits both your plants and the environment. So, let's embark on this exciting journey and discover the secrets to successful fertilization and nutrient management in no-till farming!

Understanding Soil Health and Nutrient Needs

In no-till farming, maintaining soil health and meeting the nutrient needs of plants are crucial for a thriving garden. A deep understanding of soil health and nutrient requirements is essential for gardeners to make informed decisions about fertilization and nutrient management. In this section, we will explore the key components of soil health, the primary nutrients required by plants, and how to assess the nutrient needs of your no-till garden.

Soil health is a complex and dynamic concept, encompassing the soil's physical, chemical, and biological properties. Good structure, adequate moisture and aeration, balanced pH levels, and a diverse population of microorganisms are the characteristics of healthy soil. These factors work together to create a fertile environment that supports robust plant growth and development.

Plants require a range of nutrients to grow and thrive, which can be broadly categorized into macronutrients and micronutrients. Macronutrients are needed in larger quantities and include nitrogen (N), phosphorus (P), and potassium (K). Nitrogen is essential for leafy growth and the production of proteins, while phosphorus supports root development, flowering, and fruiting. Potassium is crucial for plant health, including disease resistance and water regulation.

Micronutrients, on the other hand, are needed in smaller quantities but are no less important for plant health. Some essential micronutrients include calcium, magnesium, sulfur, iron, and zinc. These nutrients are vital in various plant processes, such as enzyme activation, photosynthesis, and cell wall formation.

To effectively manage the nutrient needs of your no-till garden, it is essential to regularly assess the soil's health and fertility. Soil testing is a valuable tool for determining your garden's nutrient levels and pH balance. By analyzing the soil test results, you can identify any nutrient deficiencies or imbalances and make informed decisions about the types and amounts of fertilizers and amendments to apply.

In addition to soil testing, observing your plants closely for signs of nutrient deficiencies is crucial. Some common symptoms include yellowing leaves, stunted growth, and poor flowering or fruiting. Addressing these issues promptly ensures your plants receive the nutrients they need to grow and thrive.

In conclusion, understanding soil health and nutrient needs is fundamental to successful no-till farming. By maintaining healthy soil, providing the necessary nutrients, and monitoring your garden's fertility, you can create a sustainable and productive no-till garden that will reward you with bountiful harvests for years to come.

Organic Fertilizers and Amendments for No-Till Gardens

In this section, we will delve into the world of organic fertilizers and amendments that can be used to enhance the fertility of your no-till garden. By choosing organic options, you are not only providing essential nutrients for your plants but also contributing to the overall health of the soil and the environment.

Organic fertilizers are derived from natural sources such as plants, animals, and minerals. They release nutrients slowly, allowing plants to absorb them at a steady pace. This gradual release of nutrients helps prevent nutrient leaching and runoff, which can harm the environment. Some popular organic fertilizers and amendments for no-till gardens include:

Compost: Compost is a nutrient-rich, organic material created from decomposed plant and animal matter. It is an excellent soil amendment that improves soil structure, water retention, and nutrient availability. Incorporate compost into your no-till garden by spreading it on the soil surface or adding it to planting holes when transplanting seedlings.

Animal Manures: Well-aged and composted animal manures, such as those from cows, horses, chickens, and rabbits, are a valuable source of nutrients for your no-till garden. They provide nitrogen, phosphorus, potassium, and beneficial microorganisms that help break down organic matter and improve soil health. Be sure to use well-composted manure to avoid introducing pathogens or burning your plants with excessive nitrogen.

Vermicompost: Also known as worm castings, vermicompost is the nutrient-rich byproduct of the decomposition process performed by earthworms. It is an excellent source of plant-available nutrients and beneficial microorganisms. Vermicompost can be added to your no-till garden by top-dressing the soil or mixing it into the planting holes.

Green Manures and Cover Crops: Green manures are plants grown specifically to be incorporated into the soil to improve fertility. Cover crops, on the other hand, are grown to protect the soil from erosion and nutrient leaching. Both green manures and cover crops can be used in no-till gardens by cutting them down and leaving them on the soil surface to decompose. This practice adds organic matter, nitrogen, and other nutrients to the soil while suppressing weeds and preventing erosion.

Organic Pelleted Fertilizers: These fertilizers are made from natural ingredients such as alfalfa meal, bone meal, feather meal, and kelp meal. They are available in a slow-release, pelleted form that can be easily applied to your no-till garden. Simply sprinkle the pellets around your plants or mix them into the soil surface.

Liquid Organic Fertilizers: Liquid organic fertilizers, such as fish emulsion, seaweed extract, and compost tea, provide a quick boost of nutrients to your plants. They can be applied to the soil or used as a foliar spray to provide nutrients directly to the plant leaves. Be sure to follow the manufacturer's instructions for proper application rates and frequencies.

Incorporating these organic fertilizers and amendments into your no-till garden provides essential nutrients for your plants and promotes a healthy, sustainable ecosystem. In the next section, we will discuss efficient application techniques for fertilizers in no-till farming.

Efficient Application Techniques for Fertilizers in No-Till Farming

In no-till farming, the application of fertilizers and amendments plays a crucial role in maintaining soil health and ensuring optimal plant growth. However, it's essential to apply these inputs efficiently to minimize waste, reduce costs, and prevent environmental issues such as nutrient runoff. In this section, we will explore several effective techniques for applying fertilizers in no-till farming systems.

Broadcast Application: This method involves evenly distributing granular or powdered fertilizers over the soil surface. It is a simple and cost-effective technique suitable for large areas and can be done manually or with the help of a spreader. However, it may not be the most efficient method for targeted nutrient delivery, as it can lead to nutrient losses due to runoff or volatilization.

Band Application: This technique involves applying fertilizers in narrow bands or strips alongside the crop rows. This targeted approach ensures that nutrients are concentrated in the root zone, leading to better nutrient uptake and reduced losses. Band application can be made using specialized equipment or by hand. It is particularly useful for applying phosphorus and potassium fertilizers, which are less mobile in the soil.

Side-Dressing: Side-dressing is the process of applying additional nutrients to the soil during the growing season, typically when plants are in their rapid growth phase. This method allows for the efficient use of nitrogen fertilizers, as it provides a timely nutrient boost to support plant growth and development. To side-dress your no-till garden, simply apply the fertilizer in a band or as a top dressing along the crop rows, taking care not to disturb the plant roots.

Foliar Application: This technique involves applying liquid fertilizers directly to the leaves of the plants, allowing for rapid nutrient absorption through the leaf tissue. Foliar application is particularly useful for addressing micronutrient deficiencies, as it provides a quick and targeted solution. However, it should be used as a supplementary method rather than a primary source of nutrients, as it cannot provide the same level of sustained nutrition as soil-based applications.

Compost Tea and Liquid Fertilizers: In no-till farming, the use of compost tea and liquid fertilizers can be an effective way to deliver nutrients directly to the root zone. These solutions can be applied through drip irrigation systems or as a soil drench, ensuring that nutrients are readily available for plant uptake. Additionally, compost tea can help improve soil structure and promote the growth of beneficial microorganisms, further enhancing the overall health of your no-till garden.

When implementing these application techniques in your no-till farming system, it's essential to consider factors such as the specific nutrient requirements of your crops, the timing of applications, and the potential for nutrient losses. By adopting efficient fertilizer application methods, you can ensure that your no-till garden receives the necessary nutrients for optimal growth while minimizing the environmental impact of your farming practices.

Monitoring and Adjusting Nutrient Levels for Optimal Plant Growth

In no-till farming, maintaining a healthy balance of nutrients in the soil is essential for optimal plant growth and overall garden success. Monitoring and adjusting nutrient levels is an ongoing process that requires keen observation, regular testing, and a proactive approach to ensure your plants have access to the essential nutrients they need to thrive. This section will discuss the importance of monitoring nutrient levels, the various methods you can use to assess your soil's health, and how to make adjustments to maintain optimal nutrient levels in your no-till garden.

The importance of monitoring nutrient levels cannot be overstated. A well-balanced soil ecosystem is the foundation of a successful no-till garden. By regularly monitoring the nutrient levels in your soil, you can identify any deficiencies or imbalances early on, allowing you to take corrective action before your plants suffer any adverse effects. This proactive approach ensures the health and vitality of your plants and contributes to the overall sustainability of your no-till garden.

There are several methods you can use to assess the nutrient levels in your no-till garden. Soil testing is the most accurate and reliable way to determine the nutrient content of your soil. You can either use a home soil test kit or send a soil sample to a professional laboratory for analysis. These tests will provide detailed information about the levels of essential nutrients, such as nitrogen, phosphorus, and potassium, as well as the pH level of your soil. It is recommended that you **perform soil tests at least once a year**, preferably in the spring or fall when soil conditions are most stable.

In addition to soil testing, you can also monitor the health of your plants for signs of nutrient deficiencies or imbalances. Some common symptoms include yellowing leaves, stunted growth, and poor fruit or flower production. By observing your plants closely and addressing any issues promptly, you can maintain a healthy balance of nutrients in your no-till garden.

Once you have identified any nutrient deficiencies or imbalances in your soil, you can take steps to adjust the nutrient levels accordingly. This may involve adding organic fertilizers or amendments, such as compost, manure, or green manure, to increase the levels of specific nutrients. Keep in mind that it is essential to use the appropriate type and amount of fertilizer or amendment to avoid over-fertilization, which can harm your plants and disrupt the balance of your soil ecosystem.

In some cases, you may need to adjust the pH level of your soil to improve nutrient availability. For instance, if your soil is too acidic, you can add lime to raise the pH, while adding sulfur can lower the pH of alkaline soils. It is crucial to make these adjustments gradually and monitor the effects on your soil and plants to avoid any adverse reactions.

In conclusion, monitoring and adjusting nutrient levels in your no-till garden is critical to maintaining a healthy and productive growing environment. By regularly assessing your soil's health, observing your plants for signs of nutrient deficiencies, and making the necessary adjustments, you can ensure that your no-till garden remains fertile and sustainable for years to come.

Achieving Sustainable Fertility in Your No-Till Garden

In conclusion, maintaining sustainable fertility in your no-till garden is achievable and highly rewarding. Following the principles and practices outlined in this chapter, you can create a thriving, productive garden that benefits your plants and the environment.

First and foremost, understanding the importance of soil health and nutrient needs is crucial in no-till farming. By regularly testing your soil and being aware of its composition, you can make informed decisions about the types of fertilizers and amendments required to maintain optimal fertility.

Organic fertilizers and amendments play a significant role in no-till gardens, as they provide essential nutrients and contribute to the soil's overall health. Using natural, eco-friendly options such as compost, manure, and cover crops can support the growth of beneficial microorganisms and improve soil structure, leading to healthier plants and higher yields.

Efficient application techniques for fertilizers in no-till farming are essential to ensure that your plants receive the nutrients they need without causing harm to the environment. By using methods such as side-dressing, fertigation, and foliar feeding, you can target specific nutrient deficiencies and minimize the risk of nutrient runoff and leaching.

Monitoring and adjusting nutrient levels for optimal plant growth is an ongoing process in no-till gardening. By keeping a close eye on the health and growth of your plants, you can identify any nutrient deficiencies or imbalances and take corrective action as needed. This proactive approach will help you maintain a fertile, productive garden throughout the growing season.

In summary, achieving sustainable fertility in your no-till garden is a rewarding endeavor that benefits both you and the environment. By focusing on soil health, using organic fertilizers and amendments, employing efficient application techniques, and regularly monitoring and adjusting nutrient levels, you can enjoy the many advantages of no-till farming while minimizing your ecological footprint. Happy gardening!

Chapter Summary

1. No-till farming requires a different approach to fertilization and nutrient management compared to conventional farming methods, focusing on maintaining soil health and meeting the nutrient needs of plants.
2. Soil health is a complex concept that includes physical, chemical, and biological properties, and healthy soil is essential for a thriving no-till garden.
3. Plants require a range of macronutrients and micronutrients for optimal growth, and understanding their specific needs is crucial for effective nutrient management.

4. Organic fertilizers and amendments, such as compost, animal manures, and cover crops, are well-suited for no-till gardens as they provide essential nutrients while also contributing to soil health and environmental sustainability.
5. Efficient application techniques for fertilizers in no-till farming, such as side-dressing and foliar feeding, help ensure that plants receive the necessary nutrients without causing harm to the environment.
6. Regularly monitoring nutrient levels in the soil and plants is essential for maintaining a healthy balance and addressing any deficiencies or imbalances early on.
7. Adjusting nutrient levels in the soil may involve adding organic fertilizers or amendments or adjusting the pH level to improve nutrient availability.
8. Achieving sustainable fertility in a no-till garden is possible and rewarding, benefiting both plants and the environment by focusing on soil health, using organic fertilizers and amendments, employing efficient application techniques, and regularly monitoring and adjusting nutrient levels.

9

HARVESTING AND POST-HARVEST PRACTICES FOR NO-TILL GARDENS

A lush garden bursting with vibrant vegetables of every color and shape, their leafy greens reaching towards the sky.

No-till gardening has gained immense popularity in recent years, and for a good reason. This sustainable and environmentally friendly approach to gardening not only improves soil health and reduces erosion but also promotes biodiversity and conserves water. As you've journeyed through the previous chapters, you've learned the ins and outs of setting up and maintaining a thriving no-till garden. Now, it's time to reap the rewards of your hard work and dedication.

In this chapter, we will delve into the world of harvesting and post-harvest practices for no-till gardens. Harvesting is an exciting and gratifying phase in the life cycle of your garden, as you finally get to enjoy the fruits (and vegetables) of your labor. However, it's essential to approach this stage with care and knowledge to ensure the optimal yield, quality, and longevity of your produce.

We will begin by discussing the importance of timing your harvest for the best possible results. This involves understanding the growth patterns and maturity indicators of various crops and considering factors such as weather conditions and pest control. Next, we will explore different techniques for harvesting a wide range of crops in a no-till garden, from leafy greens and root vegetables to fruits and herbs. Each type of produce requires a unique approach to ensure minimal damage and maximum freshness.

Post-harvest handling and storage play a crucial role in preserving the quality and nutrient content of your garden's bounty. We will provide you with practical tips and guidelines for properly handling, cleaning and storing your produce to maintain its freshness and flavor for as long as possible. Additionally, we will address the importance of reducing waste and utilizing crop residues in no-till gardens, which contributes to the overall sustainability of this gardening method.

Finally, we will conclude by emphasizing the significance of adopting sustainable harvesting and post-harvest practices in no-till gardening. By doing so, you will enjoy a bountiful and delicious harvest and contribute to the health of our planet and its ecosystems.

So, let's dive in and discover the best practices for harvesting and post-harvest handling in your no-till garden, ensuring that you make the most of your hard work and dedication to this sustainable approach to gardening.

Timing Your Harvest for Optimal Yield and Quality

In no-till gardening, as in any other form of agriculture, **timing is everything.** Knowing when to harvest your crops is crucial for obtaining the best possible yield and quality. In this section, we will explore the factors that determine the optimal time to harvest various crops in a no-till garden and how to recognize the signs that your plants are ready for harvest.

One of the primary factors to consider when determining the right time to harvest is the maturity of the crop. Each plant species has its own unique growth cycle, which dictates the ideal time for harvesting. As a no-till gardener, it is essential to familiarize yourself with the specific growth patterns and maturity indicators of the crops you are growing. This knowledge will enable you to make informed decisions about when to harvest each crop for maximum yield and quality.

Temperature, sunlight, and moisture levels also significantly determine the optimal time for harvesting. For instance, leafy greens and root vegetables generally prefer cooler temperatures, while fruiting crops like tomatoes and peppers thrive in warmer conditions. Therefore, paying attention to the weather and adjusting your harvest schedule accordingly can help ensure that your crops are harvested at the peak of their flavor and nutritional value.

Another critical aspect of timing your harvest is understanding the signs that your plants are ready for harvest. These signs can vary depending on the type of crop, but some common indicators include the following:

- **Leafy greens:** Harvest when leaves are tender and vibrant in color. Over-mature leaves may become tough and bitter.

- **Root vegetables:** Harvest when the tops of the roots are visible above the soil surface and the roots have reached their desired size.
- **Fruiting crops:** Harvest when fruits have reached their full color and size and are easily detached from the plant.
- **Legumes:** Harvest when pods are plump and filled with seeds, but before they become tough and fibrous.

In addition to these general guidelines, it is essential to monitor your plants closely and adjust your harvest timing based on their specific needs and conditions. Regularly inspecting your garden will allow you to identify any changes in plant health or growth patterns, which can help you make more accurate decisions about when to harvest.

In conclusion, correct harvest timing is a critical aspect of no-till gardening that can significantly impact the yield and quality of your crops. By understanding the growth patterns and maturity indicators of your plants, and closely monitoring their health and environmental conditions, you can ensure that your no-till garden produces the best possible harvest.

Techniques for Harvesting Different Crops in a No-Till Garden

In a no-till garden, the harvesting process is just as crucial as the planting and growing stages. Proper harvesting techniques ensure that you reap the fruits of your labor while maintaining the integrity of your garden's soil structure. In this section, we will explore various methods for harvesting different crops in a no-till garden, ensuring that you maximize your yield and preserve the health of your plants.

Leafy Greens: When harvesting leafy greens such as lettuce, spinach, and kale, it's essential to use a gentle touch. Instead of pulling the entire plant from the ground, simply snip or pinch off the outer leaves, leaving the central growing point intact. This method allows the plant to continue producing new leaves, providing you with a continuous harvest throughout the growing season.

Root Vegetables: For root vegetables like carrots, beets, and radishes, wait until they have reached their desired size before harvesting. To remove the vegetables from the ground, gently loosen the soil around the base of the plant with a hand fork or trowel. Then, grasp the plant firmly at the base of the leaves and pull upwards, taking care not to damage the surrounding soil structure.

Fruiting Vegetables: Tomatoes, peppers, and eggplants require a bit more finesse when harvesting. To avoid damaging the plant or fruit, use a pair of clean, sharp scissors or pruning shears to cut the stem of the fruit, leaving a small portion of the stem attached. Harvest fruiting vegetables when they are fully ripe and have reached their peak color and flavor.

Legumes: Beans and peas should be harvested when the pods are plump but not overly mature. To harvest, gently hold the pod with one hand and use the other hand to pinch or snip the stem, taking care not to damage the plant. Regular harvesting of legumes encourages the plant to continue producing more pods.

Cucurbits: Cucumbers, squash, and melons should be harvested when they reach their optimal size and color. To harvest, use a clean, sharp knife or pruning shears to cut the fruit from the vine, leaving a small portion of the stem attached. Be careful not to damage the vine or other fruits during the harvesting process.

Alliums: Onions, garlic, and leeks can be harvested when the tops begin to yellow and fall over. Gently loosen the soil around the base of the plant with a hand fork or trowel, then grasp the plant firmly at the base of the leaves and pull upwards. Allow the alliums to cure in a well-ventilated area for several days before storing.

By employing these harvesting techniques in your no-till garden, you'll enjoy a bountiful yield and preserve the health and vitality of your soil. Remember, the key to successful no-till gardening lies in working with nature, not against it. So, harvest with care and reap the rewards of your sustainable gardening efforts.

Post-Harvest Handling and Storage to Maintain Freshness and Nutrient Content

In a no-till garden, the focus is not only on growing healthy and bountiful crops but also on ensuring that the harvested produce retains its freshness and nutrient content. Proper post-harvest handling and storage play a crucial role in achieving this goal. In this section, we will explore various techniques and best practices that can help you maintain the quality of your harvest from when it leaves the garden until it reaches your table.

First and foremost, it is essential to **handle your produce with care** during the harvesting process. Be gentle when picking fruits and vegetables, as bruising or damaging them can lead to a faster deterioration of their quality. Use clean and sharp tools to make clean cuts, and avoid pulling or tearing the plants. This helps maintain the freshness of the produce and minimizes the risk of

introducing pathogens that could cause spoilage.

Once your produce has been harvested, it is important to cool it down as quickly as possible. This is because the rate of respiration, which affects the freshness and nutrient content of fruits and vegetables, is directly related to temperature. By lowering the temperature, you can slow down the respiration process and extend the shelf life of your harvest. Depending on the type of produce, you can use various cooling methods such as room cooling, forced-air cooling, or hydro cooling.

Proper storage is another critical aspect of maintaining the quality of your no-till garden harvest. Different fruits and vegetables have different storage requirements in terms of temperature, humidity, and air circulation. For instance, leafy greens and cruciferous vegetables generally prefer colder temperatures and high humidity, while root vegetables and winter squashes can be stored at slightly higher temperatures and lower humidity levels. Therefore, it is essential to research the specific storage needs of each crop and create a suitable environment for them.

When storing your produce, **keep it clean and dry**, as moisture can promote mold and bacteria growth. Additionally, avoid overcrowding your storage space, as this can lead to poor air circulation and increased spoilage. Regularly inspect your stored produce for any signs of decay or damage, and remove affected items promptly to prevent the spread of spoilage.

Finally, consider preserving your harvest through canning, freezing, or dehydrating. These techniques can help you enjoy the fruits of your no-till garden throughout the year while maintaining their nutritional value. Moreover, preserving your produce can also help reduce food waste and contribute to a more sustainable lifestyle.

In conclusion, proper post-harvest handling and storage are essential for maintaining the freshness and nutrient content of your no-till garden harvest. By following the best practices outlined in this section, you can ensure that your hard work in the garden translates into delicious and nutritious meals for you and your family.

Reducing Waste and Utilizing Crop Residues in No-Till Gardens

In a world where food waste and environmental concerns are increasingly pressing issues, no-till gardening offers a sustainable and eco-friendly approach to growing your own food. One of the key aspects of this method is the efficient use of resources, which includes reducing waste and utilizing crop residues. This section will explore various ways to minimize waste and make the most of the leftovers from your no-till garden.

Firstly, planning your garden carefully is essential to ensure you're growing the right amount of food for your needs. By doing so, you can avoid overproduction and reduce the chances of food waste. Keep track of your consumption patterns and adjust your planting schedule accordingly. This not only helps in reducing waste but also contributes to the overall health of your no-till garden by preventing overcrowding and competition for resources.

When it comes to harvesting, be mindful of the amount you pick at a time. Only harvest what you need for immediate consumption or preservation, leaving the rest of the crop to continue growing. This practice minimizes waste and ensures that you have a continuous supply of fresh produce throughout the season.

After harvesting, you'll likely be left with crop residues such as leaves, stems, and roots. Instead of discarding these materials, consider incorporating them back into your no-till garden. Crop residues serve as a valuable source of organic matter, which can be used to enrich the soil and support the growth of beneficial microorganisms. Leaving the residues on the soil surface or lightly mixing them into the top layer can enhance the soil structure and promote nutrient cycling.

In addition to using crop residues as a soil amendment, you can also repurpose them in other ways. For instance, you can use the residues as mulch to suppress weeds, conserve moisture, and regulate soil temperature. Alternatively, you can add the residues to your compost pile, where they will decompose and eventually become nutrient-rich compost for your garden.

Another way to reduce waste and utilize crop residues is by practicing crop rotation. By growing different crops in the same area over several seasons, you can help to balance the nutrient demands on the soil and minimize the build-up of pests and diseases. This reduces the need for external inputs such as fertilizers and pesticides and allows you to make the most of the available resources in your no-till garden.

In conclusion, reducing waste and utilizing crop residues are essential components of sustainable no-till gardening. By adopting these practices, you can minimize your environmental impact, save resources, and contribute to a healthier and more productive garden. Remember, ***every small effort counts*** when it comes to creating a more sustainable and eco-friendly world.

The Importance of Sustainable Harvesting and Post-Harvest Practices in No-Till Gardening

Sustainable harvesting and post-harvest practices play a crucial role in the success of no-till gardening. By implementing the techniques and strategies discussed in this chapter, you can ensure that your no-till garden thrives, producing bountiful yields of high-quality, nutrient-dense produce while minimizing waste and maintaining the health of your soil.

The journey of a no-till gardener is one of continuous learning and adaptation. As you gain experience and knowledge, you will become more adept at recognizing the optimal time to harvest your crops, ensuring you are rewarded with the best possible taste and nutritional value. By employing gentle harvesting techniques and being mindful of the needs of different crops, you can minimize damage to your plants and soil, preserving the integrity of your no-till garden.

Post-harvest handling and storage are equally important in maintaining the freshness and nutrient content of your produce. By taking care to properly clean, store, and preserve your harvest, you can extend its shelf life and enjoy the fruits of your labor for longer periods. This not only benefits you and your family but also contributes to a more sustainable food system by reducing food waste and the associated environmental impacts.

In a no-till garden, every part of the plant has value. By utilizing crop residues and reducing waste, you can further enhance the sustainability of your garden. These practices contribute to the health of your soil and provide valuable resources for composting and mulching, which in turn support the growth of future crops.

Ultimately, the success of a no-till garden is dependent on the care and attention given to each stage of the growing process, from planting to harvesting and beyond. By embracing sustainable harvesting and post-harvest practices, you are not only nurturing your garden but also contributing to a more sustainable and environmentally friendly approach to food production. As you continue on your no-till gardening journey, take pride in the knowledge that your efforts are making a **positive impact** on both your immediate surroundings and the wider world.

Chapter Summary

1. No-till gardening is a sustainable and environmentally friendly approach to gardening that improves soil health, reduces erosion, promotes biodiversity, and conserves water.
2. Timing your harvest is crucial for obtaining the best possible yield and quality. Familiarize yourself with the specific growth patterns and maturity indicators of the crops you are growing, and consider factors such as weather conditions and pest control.
3. Different crops require unique harvesting techniques to ensure minimal damage and maximum freshness. For example, leafy greens should be snipped or pinched off, while root vegetables should be gently loosened from the soil with a hand fork or trowel.
4. Proper post-harvest handling and storage play a crucial role in preserving the quality and nutrient content of your garden's bounty. Handle your produce with care, cool it down as quickly as possible, and store it according to its specific temperature, humidity, and air circulation requirements.
5. Reducing waste and utilizing crop residues in no-till gardens contribute to the overall sustainability of this gardening method. Plan your garden carefully to avoid overproduction, only harvest what you need, and incorporate crop residues back into the soil or use them as mulch or compost.
6. Practicing crop rotation can help balance nutrient demands on the soil and minimize the build-up of pests and diseases, reducing the need for external inputs such as fertilizers and pesticides.
7. Sustainable harvesting and post-harvest practices are essential for the success of no-till gardening, ensuring bountiful yields of high-quality, nutrient-dense produce while minimizing waste and maintaining soil health.

8. Embracing sustainable harvesting and post-harvest practices in no-till gardening contributes to a more sustainable and environmentally friendly food production approach, positively impacting your immediate surroundings and the wider world.

10

THE ENVIRONMENTAL AND ECONOMIC BENEFITS OF NO-TILL FARMING AT HOME

A sea turtle swimming through a plastic-filled ocean.

As the world grapples with the challenges of climate change, dwindling natural resources, and increasing demand for food, adopting sustainable practices in our daily lives has become more important than ever. One such practice that has been gaining momentum among home gardeners and small-scale farmers is no-till farming. This innovative approach to agriculture promotes **environmental stewardship** and offers numerous economic benefits for those who embrace it.

No-till farming, as the name suggests, is a method of growing crops without disturbing the soil through tillage. By leaving the soil undisturbed, this practice helps preserve its natural structure and fertility, leading to healthier and more productive gardens. This chapter will explore the various environmental and economic benefits of adopting no-till farming practices at home, from reducing soil erosion and conserving water to boosting biodiversity and lowering costs.

As you embark on this journey towards a more sustainable and profitable home garden, you will discover that no-till farming is not just a trend but a transformative approach to agriculture that can have lasting positive effects on the environment and your wallet. By understanding and implementing these practices, you will be contributing to a greener and more sustainable future for generations.

So, let's begin our exploration of the myriad benefits of no-till farming at home and learn how you can make a difference in your own backyard.

Reducing Soil Erosion and Improving Soil Health

No-till farming at home is a game-changer when it comes to reducing soil erosion and improving soil health. By eliminating the need for tilling, we can preserve the natural structure of the soil and create a more sustainable environment for our plants to thrive. In this section, we'll explore the ways by which no-till farming contributes to healthier soil and a more vibrant home garden.

First, let's talk about soil erosion. When we till the soil, we disturb its natural layers and expose it to the elements. This can lead to the loss of valuable topsoil through wind and water erosion. Topsoil is essential for plant growth, as it contains the nutrients and organic matter plants need to thrive. By adopting no-till practices, we can keep the soil intact and minimize erosion, preserving the precious topsoil for future generations of plants.

No-till farming also has a significant impact on soil health. Healthy soil is teeming with life, from microscopic bacteria and fungi to larger organisms like earthworms. These organisms are crucial in breaking down organic matter and making nutrients available to plants. Tilling the soil can disrupt these beneficial organisms, leading to a decline in soil health and fertility.

The soil structure remains undisturbed in a no-till garden, allowing these organisms to flourish. As they break down organic matter, they create nutrient-rich humus, which improves the soil's ability to hold water and nutrients. This leads to a more fertile and productive garden, with less need for chemical fertilizers and amendments.

Another advantage of no-till farming is the increased organic matter in the soil. When we leave plant residues on the surface, they decompose and add valuable organic matter to the soil. This organic matter provides nutrients for plants and helps improve the soil's structure, making it more resistant to compaction and erosion.

Furthermore, no-till farming can help to **reduce the risk of soil-borne diseases.** By not turning the soil, we can prevent the spread of pathogens that may be lurking beneath the surface. This can lead to healthier plants and a more bountiful harvest.

In summary, no-till farming at home offers numerous benefits for soil health and erosion control. We can create a more sustainable and productive garden by preserving the soil's natural structure and fostering a thriving ecosystem of beneficial organisms. Embracing no-till practices is a significant step towards a greener, healthier future for our home gardens and the environment as a whole.

Conserving Water and Enhancing Water Quality

One of the most significant advantages of adopting no-till farming at home is its ability to conserve water and enhance water quality. As water scarcity becomes an increasingly pressing issue worldwide, it's essential to explore sustainable gardening practices that help preserve this vital resource. In this section, we'll delve into how no-till farming contributes to water conservation and improved water quality in your home garden.

No-till farming is an excellent way to conserve water, as it helps retain moisture in the soil. When you till the soil, you expose it to the sun and wind, which can cause rapid evaporation of the moisture within. By leaving the soil undisturbed, no-till farming allows the ground to maintain its natural structure, which is better suited to hold water. This means that your plants will require less irrigation, ultimately reducing your water usage and saving you money on your water bill.

In addition to conserving water, no-till farming also enhances water quality. Traditional tilling methods can lead to **soil erosion**, which in turn can cause sediment and other pollutants to enter nearby water sources. This sedimentation can harm aquatic ecosystems and reduce the overall quality of the water. No-till farming helps prevent soil erosion by keeping the soil structure intact, which means that fewer pollutants will find their way into our water systems.

Moreover, no-till farming promotes the growth of beneficial microorganisms and earthworms that help break down organic matter and improve soil fertility. These organisms contribute to the formation of stable soil aggregates, which improve the soil's ability to absorb and hold water. This increased water infiltration reduces runoff, which is a significant source of water pollution. By minimizing runoff, no-till farming practices help protect our rivers, lakes, and streams from contamination.

Furthermore, no-till farming reduces the need for chemical fertilizers and pesticides, which can also contaminate water sources. By fostering a healthy soil ecosystem, no-till practices encourage the growth of robust, disease-resistant plants that require fewer chemical inputs. This benefits the environment and reduces the risk of ingesting harmful chemicals through the food we grow in our home gardens.

In conclusion, adopting no-till farming practices at home is a powerful way to conserve water and enhance water quality. By preserving soil structure, reducing erosion, and minimizing the need for chemical inputs, no-till farming contributes to a healthier, more sustainable home garden. As we face growing concerns about water scarcity and pollution, embracing no-till farming is a crucial step towards a greener, more responsible future.

Boosting Biodiversity and Supporting Pollinators

A flourishing garden is a haven for a diverse range of organisms, from the tiniest microbes to the most captivating butterflies. No-till farming at home not only nurtures the soil and conserves water, but it also plays a vital role in boosting biodiversity and supporting pollinators. This section will explore how no-till practices contribute to a thriving ecosystem in your backyard, creating a sanctuary for plants and wildlife.

To appreciate the significance of biodiversity, it's essential to understand that every organism in an ecosystem plays a crucial role in maintaining balance and harmony. A diverse ecosystem is more resilient, adaptable, and productive. No-till farming at home encourages the growth of various plant species, which in turn attracts a wide array of insects, birds, and other wildlife. Avoiding soil disturbance allows natural processes to take place, fostering a healthy and balanced ecosystem.

One of the most critical groups of organisms that benefit from no-till farming practices is pollinators. Pollinators, such as bees, butterflies, and hummingbirds, are indispensable to our food system as they help in the fertilization of plants, leading to the production of fruits, vegetables, and seeds. Unfortunately, pollinator populations have been declining globally due to habitat loss, pesticide exposure, and climate change. By adopting no-till practices at home, you can **create a safe haven** for these essential creatures.

No-till farming at home supports pollinators in several ways. First, it reduces the need for chemical inputs, such as pesticides and herbicides, which can harm pollinators. Maintaining a chemical-free environment provides a safe space for pollinators to thrive. Second, no-till practices promote the growth of cover crops and diverse plant species, offering a variety of nectar and pollen sources for pollinators. This abundance of food sources attracts and sustains a healthy pollinator population in your garden.

Moreover, no-till farming practices contribute to the creation of microhabitats, which provide shelter and breeding grounds for various organisms. For instance, undisturbed soil is home to a myriad of beneficial insects, such as ground beetles and spiders, which help control pests. Similarly, leaving plant residues on the soil surface creates a habitat for decomposers like earthworms, which break down organic matter and improve soil fertility.

In conclusion, no-till farming at home is a powerful tool for boosting biodiversity and supporting pollinators. By embracing this sustainable approach, you can create a vibrant, resilient, and productive garden that benefits both humans and wildlife. As more home gardeners adopt no-till practices, we can collectively contribute to the *preservation of our planet's precious ecosystems* and the invaluable services they provide.

Lowering Costs and Increasing Profitability for Home Gardeners

No-till farming is not only an environmentally friendly approach to home gardening, but it also offers significant economic benefits. By adopting no-till practices, home gardeners can lower their costs and increase profitability in several ways. So let's explore how no-till farming can help you save money while enjoying a bountiful harvest.

First and foremost, no-till farming eliminates the need for expensive and labor-intensive tilling equipment. Traditional tilling methods require gardeners to invest in tools such as rototillers or plows, which can be costly to purchase, maintain, and operate. By forgoing these tools, no-till gardeners can save money on equipment and fuel costs. Additionally, the time saved by not tilling can be allocated to other gardening tasks or leisure activities, making no-till farming a more efficient and enjoyable approach.

Secondly, no-till farming practices contribute to improved soil health, leading to higher crop yields. Healthy soil is rich in organic matter and nutrients, providing an ideal environment for plants to thrive. As a result, no-till gardeners often experience increased harvests without the need for costly chemical fertilizers. This not only saves money but also reduces the environmental impact of your home garden.

Water conservation is another economic benefit of no-till farming. By preserving soil structure and promoting water retention, no-till practices reduce the need for frequent irrigation. This can lead to significant savings on water bills, especially in regions with limited water resources or high water costs. Furthermore, by reducing runoff and soil erosion, no-till farming helps maintain water quality, which is essential for healthy plant growth and can save money on water treatment expenses.

No-till farming also supports beneficial insects and pollinators, which play a crucial role in the success of your home garden. By providing a habitat for these essential creatures, no-till practices can increase pollination and natural pest control in your garden, reducing the need for costly and harmful pesticides. This not only saves money but also contributes to a healthier and more sustainable garden ecosystem.

Lastly, no-till farming can increase the marketability of your home-grown produce. As consumers become more environmentally conscious, there is a growing demand for sustainably grown fruits and vegetables. By adopting no-till practices, you can cater to this demand and potentially earn a premium for your produce at local farmers' markets or through direct sales to neighbors and friends.

In conclusion, no-till farming offers a multitude of economic benefits for home gardeners. By reducing costs associated with equipment, fertilizers, water, and pesticides and by increasing crop yields and marketability, no-till practices can transform your home garden into a more profitable and sustainable endeavor. So, embrace the no-till revolution and reap the rewards of a healthier, more cost-effective, and environmentally friendly home garden.

Transforming Home Gardening with No-Till Practices

As we have explored throughout this chapter, the environmental and economic benefits of no-till farming at home are numerous and significant. By adopting these sustainable practices, home gardeners can play a crucial role in preserving our planet's resources and promoting a healthier ecosystem. In this concluding section, we will recap the key advantages of no-till farming and discuss how it can revolutionize the way we approach home gardening.

First and foremost, no-till farming is a powerful tool in combating soil erosion and improving soil health. By leaving the soil undisturbed, we allow it to maintain its structure and retain vital nutrients. This, in turn, leads to a more fertile and productive garden capable of supporting a diverse range of plant life. Furthermore, the enhanced soil structure allows for better water infiltration and retention, reducing the need for irrigation and conserving our precious water resources.

The benefits of no-till farming extend beyond the soil, as it also contributes to cleaner waterways and improved water quality. In addition, by minimizing soil disturbance, we reduce the risk of nutrient runoff and sedimentation, which can have detrimental effects on aquatic ecosystems. This is particularly important for home gardeners, as our collective efforts can significantly impact the health of our local watersheds.

No-till practices also promote biodiversity and support pollinators, which are essential for

maintaining a thriving ecosystem. By providing a natural habitat for various organisms, we encourage a balanced and resilient environment that can better withstand pests, diseases, and other challenges. This not only benefits our gardens but also contributes to the overall health and well-being of our planet.

From an economic standpoint, no-till farming offers numerous advantages for home gardeners. By reducing the need for tilling equipment, fertilizers, and pesticides, we can lower our gardening expenses and increase profitability. Moreover, the improved soil health and water conservation associated with no-till practices can lead to higher crop yields and better-quality produce, further enhancing the economic benefits of this approach.

In conclusion, no-till farming has the potential to transform the way we approach home gardening, offering a more sustainable and environmentally friendly alternative to traditional practices. By embracing no-till methods, we can contribute to a healthier planet while reaping the rewards of a more productive and cost-effective garden. As we move towards a future marked by increasing environmental challenges, it is more important than ever for home gardeners to adopt practices that support the well-being of our planet and its inhabitants. No-till farming is a powerful step in that direction, and its widespread adoption can pave the way for a ***greener and more sustainable future.***

Chapter Summary

1. No-till farming is a sustainable agricultural practice that involves growing crops without disturbing the soil, preserving its natural structure and fertility.
2. Adopting no-till farming at home can significantly reduce soil erosion and improve soil health, leading to a more productive and fertile garden.
3. No-till practices contribute to water conservation and enhanced water quality by preserving soil structure, reducing erosion, and minimizing the need for chemical inputs.

4. By fostering a healthy soil ecosystem and promoting diverse plant species, no-till farming boosts biodiversity and supports pollinators, which are essential for maintaining a thriving ecosystem.
5. No-till farming eliminates the need for expensive and labor-intensive tilling equipment, reducing costs for home gardeners.
6. Improved soil health, water conservation, and support for beneficial insects and pollinators can lead to higher crop yields, and better-quality produce, increasing the profitability of home gardens.
7. No-till farming practices can cater to the growing demand for sustainably grown fruits and vegetables, potentially earning a premium for homegrown produce.
8. Embracing no-till farming at home is a powerful step towards a greener, more sustainable future, offering environmental and economic benefits that can transform the way we approach home gardening.

EMBRACING A SUSTAINABLE FUTURE WITH NO-TILL FARMING

As we reach the conclusion of this enlightening exploration into the world of no-till farming at home, it is essential to pause and reflect on the journey we have taken together. From understanding the basics of no-till farming to delving into its numerous benefits, we have uncovered the transformative potential of this sustainable practice for our gardens, communities, and planet.

The no-till journey is not merely about changing the way we cultivate our home gardens; it is about embracing a new mindset that values the long-term health of our soil, our environment, and our well-being. By adopting no-till farming practices, we are making a conscious decision to break away from traditional, resource-intensive methods that have contributed to soil degradation and environmental harm. Instead, we are choosing a path that nurtures the natural processes of the earth, allowing it to flourish and sustain life for generations to come.

Throughout this book, we have delved into the science behind no-till farming, the techniques that make it successful, and the incredible benefits it offers. We have learned how no-till farming can lead to a greener tomorrow by reducing erosion, increasing soil fertility, and promoting biodiversity. We have also discovered how these sustainable practices can strengthen our communities by encouraging local food production, fostering collaboration, and reducing our reliance on harmful chemicals and synthetic fertilizers.

Moreover, we have explored the economic benefits of no-till farming at home, revealing how this approach can save us money and resources in the long run by minimizing the need for expensive inputs and labor-intensive practices. And perhaps most importantly, we have recognized the power of no-till farming to inspire future generations to adopt sustainable agricultural practices, ensuring a healthier and more resilient world for all.

As we move forward from this point, let us carry the knowledge and insights gained from our no-till journey into our everyday lives. By doing so, we can continue the no-till legacy, fostering a sustainable future that benefits not only ourselves but the generations that follow.

The Environmental Impact: A Greener Tomorrow

As we stand on the precipice of a new era in agriculture, we must recognize the profound environmental impact that no-till farming at home can have on our planet. By embracing this sustainable practice, we are not only nurturing our gardens and crops but also actively contributing to a greener tomorrow. In this section, we will delve into the various ways in which no-till farming positively affects the environment and how it plays a crucial role in mitigating climate change.

First and foremost, no-till farming significantly reduces soil erosion. Traditional tilling methods disrupt the soil structure, making it more susceptible to being washed or blown away. By leaving the soil undisturbed, no-till farming allows the natural ecosystem to thrive, fostering the growth of beneficial microorganisms and earthworms that help maintain soil health. This, in turn, leads to improved water retention and nutrient availability, which ultimately results in healthier and more resilient plants.

Another noteworthy environmental benefit of no-till farming is its ability to sequester carbon in the soil. As we know, excessive carbon dioxide in the atmosphere is a leading cause of global warming. By adopting no-till practices, we can help combat climate change by storing more carbon in the soil, thereby reducing greenhouse gas emissions. Moreover, no-till farming requires less energy and fuel consumption, as there is no need for heavy machinery to till the soil. This further contributes to a **decrease in carbon emissions** and a smaller environmental footprint.

No-till farming also plays a vital role in preserving biodiversity. By not disturbing the soil, we allow the natural balance of insects, microorganisms, and other wildlife to remain intact. This promotes a healthy ecosystem that supports a wide variety of species, which is essential for the overall well-being of our planet. Additionally, no-till farming can help reduce the need for chemical fertilizers and pesticides, as healthier soil and plants are more resistant to pests and diseases. This, in turn, leads to cleaner water sources and a safer environment for both humans and wildlife.

In conclusion, the environmental impact of no-till farming at home cannot be overstated. By adopting this sustainable practice, we are actively contributing to a greener and healthier planet for ourselves and future generations. The benefits of no-till farming extend far beyond our backyards, as they play a crucial role in mitigating climate change, preserving biodiversity, and promoting a more sustainable future. As we continue our no-till journey, let us take pride in knowing that our efforts are making a tangible difference in the world around us.

Strengthening Our Communities Through Sustainable Practices

As we journey towards a greener and more sustainable future, it is essential to recognize the role that no-till farming at home plays in strengthening our communities. By adopting sustainable practices, we contribute to our environment's well-being and foster a sense of unity and shared responsibility among community members. In this section, we will explore how no-till farming can bring people together and create a lasting impact on the social fabric of our neighborhoods.

First and foremost, no-till farming at home encourages a sense of collective ownership and pride in our communities. When individuals take the initiative to cultivate their gardens and landscapes using sustainable methods, they become active participants in the stewardship of their local environment. This shared commitment to preserving the earth's resources can inspire neighbors to collaborate and exchange ideas, fostering a sense of camaraderie and interconnectedness.

Moreover, no-till farming can catalyze community-based initiatives and projects. For instance, neighbors can come together to establish communal gardens, where they can practice no-till farming techniques and grow fresh produce for the entire community. Such initiatives promote sustainable agriculture and provide opportunities for social interaction, skill-sharing, and the development of strong support networks among community members.

In addition, no-till farming at home can contribute to the local economy by reducing the dependence on commercially grown produce. By growing their fruits and vegetables, individuals can save money on groceries and support local businesses by purchasing other goods and services. Furthermore, the surplus produce from no-till gardens can be shared with neighbors or donated to local food banks, helping to address food insecurity and promote a culture of generosity and mutual support.

Educational opportunities also abound when communities embrace no-till farming practices. Schools can incorporate sustainable agriculture into their curricula, teaching students the importance of soil health, water conservation, and biodiversity. By engaging in hands-on learning experiences, children can develop a deep appreciation for the environment and the role they play in preserving it for future generations.

In conclusion, adopting no-till farming at home can significantly strengthen our communities by fostering a sense of unity, promoting collaboration, and contributing to the local economy. By embracing these sustainable practices, we can work together to create a brighter, greener future for ourselves and the generations to come.

The Economic Benefits of No-Till Farming at Home

As we reach the end of our no-till farming journey, it's essential to consider the economic benefits that this sustainable practice brings to our homes and communities. No-till farming at home contributes to a greener environment and offers many financial advantages that make it an appealing choice for gardeners and small-scale farmers alike.

First and foremost, no-till farming saves time and labor. Traditional tilling methods require a significant amount of physical effort and can be time-consuming, especially for those with larger gardens or plots. By eliminating the need to till the soil, no-till farming allows individuals to allocate their time and energy to other essential tasks, such as planting, harvesting, and maintaining their gardens.

In addition to saving time and labor, no-till farming reduces the need for expensive equipment and machinery. Tilling the soil typically requires the use of tractors, plows, and other specialized tools, which can be costly to purchase, maintain, and repair. By adopting a no-till approach, gardeners and small-scale farmers can avoid these expenses and invest their resources in other areas, such as acquiring high-quality seeds, compost, or irrigation systems.

Another economic benefit of no-till farming at home is the potential for increased crop yields. Healthy, well-structured soil is crucial for plant growth, and no-till farming promotes soil health by preserving its natural structure and fostering a thriving ecosystem of beneficial microorganisms. As a result, no-till gardens and farms often experience higher crop yields, which can translate into increased profits for those who sell their produce or a more abundant harvest for those who grow food for personal consumption.

Furthermore, no-till farming can reduce input costs, as it often requires fewer fertilizers and pesticides. The improved soil structure and increased organic matter in no-till systems promote nutrient availability and water retention, reducing the need for synthetic fertilizers. Additionally, the diverse ecosystem of microorganisms and insects in no-till soil can help control pests and diseases, minimizing the reliance on chemical pesticides.

Lastly, no-till farming at home can contribute to long-term financial stability by safeguarding the productivity of the land. Soil erosion and degradation are significant concerns in conventional farming practices, which can lead to declining crop yields and, ultimately, the need to invest in costly soil remediation measures. By preserving soil health and preventing erosion, no-till farming ensures that the land remains fertile and productive for future generations.

In conclusion, the economic benefits of no-till farming at home are numerous and far-reaching. By saving time and labor, reducing equipment and input costs, increasing crop yields, and promoting long-term land productivity, no-till farming offers a sustainable and financially viable alternative to traditional agricultural practices. As we embrace this innovative approach, we are investing not only in the health of our environment but also in the prosperity of our homes and communities.

Inspiring Future Generations to Adopt No-Till Farming

As we look toward the future, we must consider the impact of our actions on the generations to come. Our decisions today will have lasting consequences on our environment's health, our communities' stability, and our planet's overall well-being. One of the most effective ways to ensure a sustainable future is by inspiring the next generation to adopt no-till farming practices.

Educating our children and young adults about the benefits of no-till farming is crucial in fostering a sense of environmental responsibility and stewardship. By incorporating lessons on sustainable agriculture into school curriculums and community programs, we can empower our youth with the knowledge and tools they need to make informed decisions about the way they cultivate and consume food.

In addition to formal education, we can also inspire future generations by leading by example. By practicing no-till farming at home and sharing our experiences with others, we can demonstrate the tangible benefits of this approach and encourage others to follow suit. This can be as simple as starting a conversation with a neighbor about your no-till garden or as ambitious as organizing a community workshop on sustainable agriculture techniques.

Moreover, we must also recognize the importance of innovation and technology in advancing no-till farming practices. By supporting research and development in this field, we can help ensure that future generations have access to cutting-edge tools and techniques that make no-till farming even more efficient and effective. This includes investing in soil health research, developing new crop varieties that thrive in no-till systems, and exploring new ways to manage pests and diseases without resorting to harmful chemicals.

Finally, we must create a supportive environment for young farmers interested in adopting no-till practices. This includes advocating for policies and incentives that make it easier for new farmers to access land, capital, and resources, as well as fostering a sense of community among like-minded individuals who can share their experiences and knowledge.

By taking these steps, we can inspire future generations to embrace no-till farming as a viable and sustainable alternative to conventional agriculture. In doing so, we will not only help to protect our environment and strengthen our communities but also ensure a brighter and more sustainable future for all.

Continuing the No-Till Legacy for a Sustainable Future

As we reach the end of our journey exploring the world of no-till farming at home, it is essential to take a moment to reflect on the importance of this sustainable practice and the legacy we are leaving for future generations. The decision to adopt no-till farming is not just a choice we make for ourselves, but it is also a commitment to the well-being of our planet and the prosperity of our communities.

By embracing no-till farming, we are taking a stand against the destructive practices that have plagued our environment for far too long. Soil erosion, water pollution, and biodiversity loss are just a few of the issues that no-till farming can help mitigate. By choosing to nurture our soil and protect our ecosystems, we contribute to a **greener**

tomorrow and ensure that our planet remains habitable for future generations.

Furthermore, no-till farming at home strengthens our communities by promoting a sense of shared responsibility for the environment. When we work together to implement sustainable practices, we foster a sense of unity and collaboration that transcends individual interests. This collective effort not only benefits our immediate surroundings but also serves as an inspiration for other communities to follow suit.

The economic benefits of no-till farming at home cannot be overlooked. By reducing the need for costly chemical inputs and heavy machinery, no-till farming allows us to invest in other areas of our lives, such as education, healthcare, and local businesses. This financial freedom can lead to a more prosperous and stable community, further reinforcing the importance of sustainable practices.

As we pass on the knowledge and skills of no-till farming to future generations, we are equipping them with the tools they need to continue the legacy of sustainable agriculture. In addition, by instilling a sense of environmental stewardship in our children, we are fostering a future where the health of our planet is a top priority. This, in turn, will ensure that the benefits of no-till farming continue to be felt for years to come.

In conclusion, adopting no-till farming at home is not just a personal choice but a powerful statement of our commitment to a sustainable future. By embracing this practice, we are taking a stand for the environment, our communities, and future generations. The legacy of no-till farming is one of hope, resilience, and prosperity – a legacy we can all be proud to be a part of. So, let us continue to nurture our soil, protect our ecosystems, and inspire those around us to join us in this journey toward a more sustainable world.

EPILOGUE

Congratulations on finishing *"Rainwater Harvesting and No-Till Farming"* This book has provided valuable insights and practical guidance on two essential practices that can help reduce your environmental footprint and contribute to a healthier planet.

By now, you have learned about the importance of rainwater harvesting and no-till farming and how they can benefit your home and garden and the environment as a whole. These practices are just two examples of the many ways we can make a positive impact on the world around us.

But it doesn't end here. We can always do more to create a more sustainable and eco-friendly future. Every little step counts, whether it's reducing our energy consumption, recycling, composting, or supporting local farmers and sustainable businesses. Together, we can make a difference.

This book has inspired you to take action and change your life. Remember, even small steps can have a significant impact. So, take what you've learned here and keep moving towards a greener future.

Thank you for choosing this book and for your commitment to sustainability. I wish you all the best in your eco-friendly journey!

ABOUT THE AUTHOR

Michael Barton is an expert in sustainability and regenerative agriculture with over a decade of experience. He holds a degree in Environmental Studies and has received advanced training in the principles and practices of regenerative agriculture.

Throughout his career, Michael has worked with farmers, policymakers, and environmentalists to promote sustainable farming practices and advocate for the adoption of regenerative agriculture. He has collaborated with organizations worldwide to advance sustainable agriculture and food systems.

In his free time, Michael enjoys hiking and spending time with his family.

Made in the USA
Monee, IL
04 November 2026

67383010R00154